How I Got This Idea?

Lingaraj Patnaik

Copyright: Lingaraj Patnaik

Cover Design: Suman Patnaik, Alok Jyoti Nanda

ISBN-13: 978-1499534245

ISBN-10: 1499534248

This book is printed on recycled paper. No trees were cut for the production of this book. All rights reserved. No part of this publication may be reproduced, stored in a retrieval system, or transmitted in any form or by any means electronic, mechanical, photocopying, recording or otherwise without the prior permission of the author. The views expressed in this book are those of the author. The publisher is not in any way responsible for the views expressed in this book.

This book is placed in the hands of

Dr. G Rajsekhar, my teacher

Contents

	Preface	7
1.	Vertical Pedal Bicycle	9
2.	Transverse Wave Propulsion in Fish and Birds	12
3.	Building Wound-Core Transformer	21
4.	Solution to Nuisance Tripping of Plant Lightning from Fault in Appliances	25
5.	Dogma, Laws of Physics and the Baconian Filter	28
6.	Derivation of the Three Laws of Newton from the Law of Conservation of Energy	38
7.	A Pressure Model for the Inverse Square Law	42
8.	Reduction of the Set of Four Electromagnetic Equations of Maxwell to a Set of Two Equations	44
9.	SECOND PAPER, Maxwell's Equations	47
10.	The Questionable Freedom of Fixing the (Ludwig) Lorenz Gauge	52
11.	A Conceptual Error in the Calculation of Length Contraction in Special Relativity	56
12.	Defining 'Scale' of Non-Singular Matrices	76
13.	A New Approach to Analyticity in Complex Analysis	79

14. Inadvisability of Import of Vector Concepts from the Real to the Complex Analysis	83
15. Review of the Complex Potential	88
16. Algebraic Solution of Chemical Equations	92
17. Sustainable Development, Entropy, Madness and the Atom Bomb	97
18. An Accounting Framework for Energy Conservation	107
19. Tyranny of Errors	114
20. Time Estimation of a Job	118
21. Industrial Organization and Pump–Pipeline Network	123
22. $P = Q \times R$ and an Industrial Organization	126
23. A Structural Limitation of Language: One Dimensionality	130
Certificates	134
Index	144

Preface

THIS BOOK CONTAINS ideas which occurred to me since my boyhood. Under each topic I have stated how I got the basic idea. Some of these ideas are:

- The vertical pedal bicycle. This idea occurred to me in my high school. This was demonstrated at IIT-BHU.

- The Transverse wave propulsion in fish. This occurred to me at Tampara, a beautiful lake near to my home at Chatrapur, Odisha. The lake is, presently, being developed as a tourist attraction by the Government of Odisha. I also demonstrated pumping based on this idea.

- I could derive all the three laws of Newton from the law of conservation of energy. One may derive the first 'law' of Newton from the second law, $F = ma$, for the special case, $F = 0$.

- Compared with fluid dynamics, where a pressure is the motive and flow is the movement, in electricity, however, both the motive (flux) and the movement (current) are described as flows. I found a pressure model for the inverse square law at IIT-BHU.

- Perhaps while moving around the beautiful hills of Parvatipuram at Nagercoil, I wondered why we should have four equations of Maxwell to deal with only two fields: the electric and the magnetic? I could reduce the system of four equations to a set of two in virtually no time only to find that others had achieved the same result earlier.

- Looking in detail at the (Hendrik) Lorentz transformations, I noticed a conceptual error in the calculation of length contraction in special relativity.

- Once in every five years or so I would visit the Cauchy-Riemann conditions for analyticity of complex functions. I felt the text book method of mechanical, sequential movements on the complex plane somewhat unsatisfactory. I discovered a new approach to the Cauchy-Riemann conditions from the concept of 'exactness' and presented a paper in a national conference on mathematics, organized by Odisha Mathematical Society and sponsored by CSIR.

- *Sustainable development as it is being propagated today resembles "eating a cake and having it too (for the future generation)", a contradiction in terms.* The heat energy that is raining down upon the earth surface from the tropopause, 10 km above, every hour because of anthropogenic activities is some 16,000 times the heat content of the atom bomb that was dropped on Hiroshima, a powerful symbol of mass destruction in modern times. A new unit "HABE" is fashioned around the name 'Hiroshima' and is proposed to measure the quantities of heat involved in global warming.

- Organizations can be modeled after pump-pipe line network where flow of work resembles the flow of fluid. This will help one to hunt out various resistances to the flow of work in an organization.

- An entire book can be written down in a single geometrical straight line; this represents a feature of one dimensionality in descriptive language. This is examined.

There are more.

Quite likely many other people, too, have worked on the same ideas and must not be deprived of their claims of precedence where evident. Some I have mentioned in the text.

I hope the reader shall enjoy the book.

Lingaraj Patnaik,

Vertical Pedal Bicycle

How I Got This Idea

Late in my boyhood I came across an article in Science Today, a popular science magazine in India, descrbing a vertical pedal bicycle invented by a student from an IIT. There were two freewheels on either side of the rear wheel. A chain ran on each freewheel. A pedal hung from each chain on a pulley on either side of the bicycle. After being pushed down, the pedals were restored to their top positions using springs.

I had a hunch. A rope hung on a pulley will have one end pulled up when the other is pulled down. Two pieces of chain may be taken, one on each of the two freewheels, fixed on either side of the rear wheel. These two pieces of chain may be connected to three pieces of rope running on three pulleys. The two pieces of chain and the three pieces of rope may be connected alternatingly to make a single length of rope-chain-rope-chain-rope. This arrangement will pull up a pedal when the other pedal is being pushed down. Now no springs are required to restore the pedals to their top positions after being pushed down. Look at the drawing in the following page.

The conventional bicycle with pedals moving in circles is, for most of the circular path, not adapted to utilise the nearly vertical force applied on the pedal. The vertical pedal system seeks to improve the situation by a system of pedals moving approximately vertically and parallel to the applied force.

Dr. G. Rajsekhar, my teacher at IIT-BHU financed this project.

Conventional Pedal System

THE PEDALS in the conventional bicycle move in circles. Only the component of the force perpendicular to the crank shaft helps transfer energy. The component parallel to the crank shaft cannot transfer energy as the crank shaft is a rigid body; there cannot be a movement parallel to the rigid crank shaft (note: energy transferred *or* work done = force * movement along the direction of force).

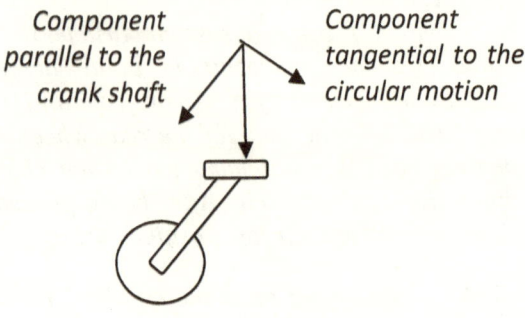

CONVENTIONAL BICYCLE

Vertical Pedal System

The rear wheel is provided with two freewheels, one on either side. An iron rope-chain-string length is arranged suitably to move on the two freewheels and three pullys and is connected to two pedals at either end. When the pedals are pushed alternately down by both feet, the energy is transferred to the rear wheel of the bicycle via the two freewheels stated above. Certificates are displayed at the end of this book.

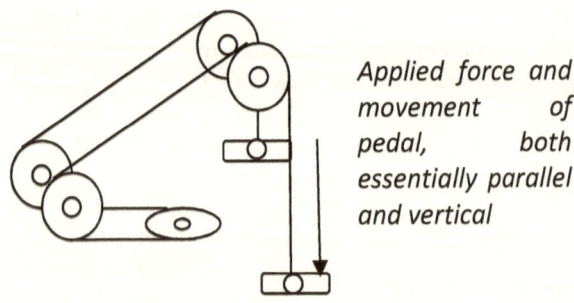

Applied force and movement of pedal, both essentially parallel and vertical

VERTICAL PEDAL BICYCLE

Finally Speaking

Finally speaking, I feel that the conventional bicycle with pedals moving in circles is a compact arrangement, a very good engineering requirement and is far better than the vertical pedal system. In case of the conventional pedal system, the vigorous pushing is done, instinctively, where the crank shaft is fairly parallel to the ground; both the direction of the push of foot as well as the tangential movement of pedal are downward, parallel to each other and quite efficient.

Transverse Wave Propulsion in Fish and Birds

How I Got This Idea

Near to my home at Chatrapur, Odisha, there is a beautiful lake, Tampara, which I used to frequent and boat across. Once I was intently watching fish trying to figure out how they swam. I ruled out the flimsy fins as propulsion apparatus. Fins may be useful for ensuring stability and control but not for providing propulsion. I observed, every time a fish starts moving, the movement starts with a side wise shake of its head and a subsequent wave in its body which proceeds to the tail. I intuited that the wave profile, while it proceeded from head to tail, pushed the surrounding water backwards. By Newton's third law of motion, the reaction from the surrounding water pushed the fish forward. Later, I took my sister's school box and put some water and a small sheatfish in to it and closely watched the fish swim. After perhaps an hour, I was convinced that I was correct. Subsequently, referring to the Encyclopedia Britannica, I found the very same concept was described to explain the motion of fish.

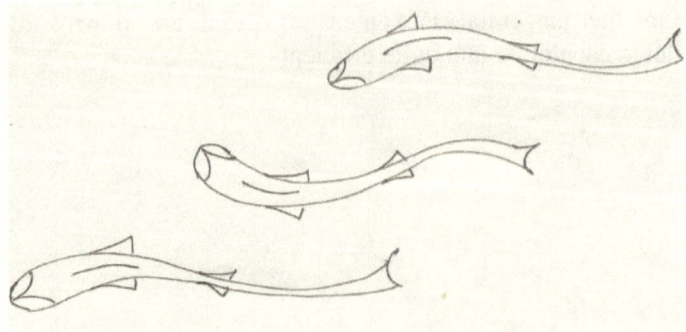

I felt I can simulate the fish mechanically. This I did in the year 1980 in the wave-boat described below. Subsequently I felt, I can create a water pump by encasing the propeller as an impeller. This I demonstrated in 1988 in a wave-pump described later on.

Still later, watching cranes, flying alongside while I was on trains slowly moving, I felt the same principle was at work in the wings of the birds; the only added complication was that the wings must also provide lift from the Bernoulli's principle (in aircrafts, however, the wings provide only lift). I demonstrated the principle of the wave-propeller in air in the year 1993. This is described at the end.

Transverse wave propulsion is fundamental to the propulsion of both fish and birds.

Chronological Development

Wave Boat, Modex- 80, Year 1980

IN THE YEAR 1980 the *Wave Boat* was exhibited in the Model Exhibition *MODEX- 80* in the IIT, Benaras Hindu University. The idea was counter intuitive considering that a cork floating in water will only bob up and down when a wave passes it by. It was a surprise to many including my professors one of whom laughed with glee when the Wave Boat moved in water.

A rubber cloth, held taut by springs, was oscillated up and down by a flywheel-crankshaft mechanism driven by a small motor. Transverse waves were created at the leading edge of the rubber sheet and proceed to the rear edge of the rubber sheet. The water surrounding the rubber sheet was pushed backwards by the profile of the transverse waves. By Newton's 3rd law of motion the boat moved forward. The body of the Wave Boat was in the shape of a pontoon.

The basic idea had been borrowed from fish. After watching carefully fish swimming in a lake I concluded that the fins, being flimsy, were at most useful for stability and control purpose, but not for the purpose of delivering propulsive power. Noticing that there is undulating movement of the body of the fish, starting at its head and proceeding to its tail every time the fish started to swim, suspicion dawned that the wave motion is indeed the fundamental mechanism of fish propulsion. That began the Wave Boat experiment. It won a prize among *Advanced Models* category.

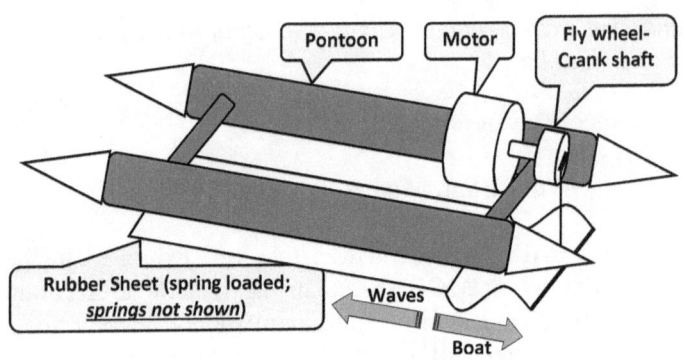

Schematic Diagram, Wave Boat (Propeller)

Wave Boat (propeller), Model Exhibition- 1980, IIT-BHU

Wave Pump, Year 1988

A pump is where the propeller is called an impeller and is confined in a casing that is held stationary. Approximately 6 gallons per minute water issued out of the discharge pipe in a smooth flow. The maximum head at no discharge was 3 meters. The impeller was oscillated about a fixed axle something like a hand fan. This was a practical engineering adaptation of the ideal transverse wave motion.

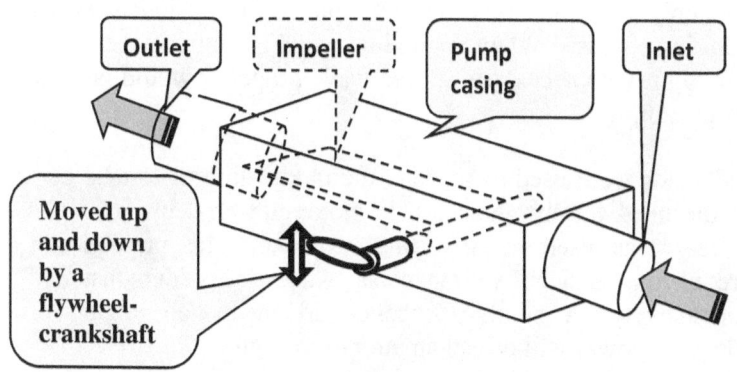

Schematic Diagram, Wave Pump

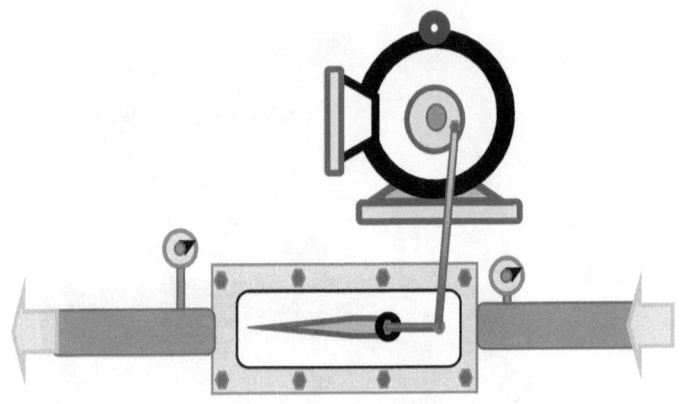

Artist's Impression of the Wave Pump

Wave Propeller in Air, Year 2003

There was no actual flying. *Only the propulsion from transverse waves was demonstrated for the air propeller.* The experiment was similar in concept to the testing of propulsion of a jet engine or a rocket engine on a test bed.

In birds the wings provide both lift as well as propulsion. In fixed wing aircrafts the wings provide lift whereas propeller or jet engine provides propulsion. The purpose of the experiment was only to demonstrate the principle of propulsion in air by transverse waves. When applied to actual aircrafts, fixed wings would provide necessary lift and wave propeller would provide propulsion.

Threads were used to suspend the model aircraft to take care of the need for lift. A small dc motor powered by 1.5 Volts battery was used to drive the propeller. The principle of propulsion in air by transverse waves was demonstrated conclusively. A similar propeller driven by an engine of adequate power will propel an aircraft to flight.

The propeller was oscillated about a fixed axle something like a hand fan. This was a practical adaptation of the ideal transverse wave motion just as in the Wave pump of 1988.

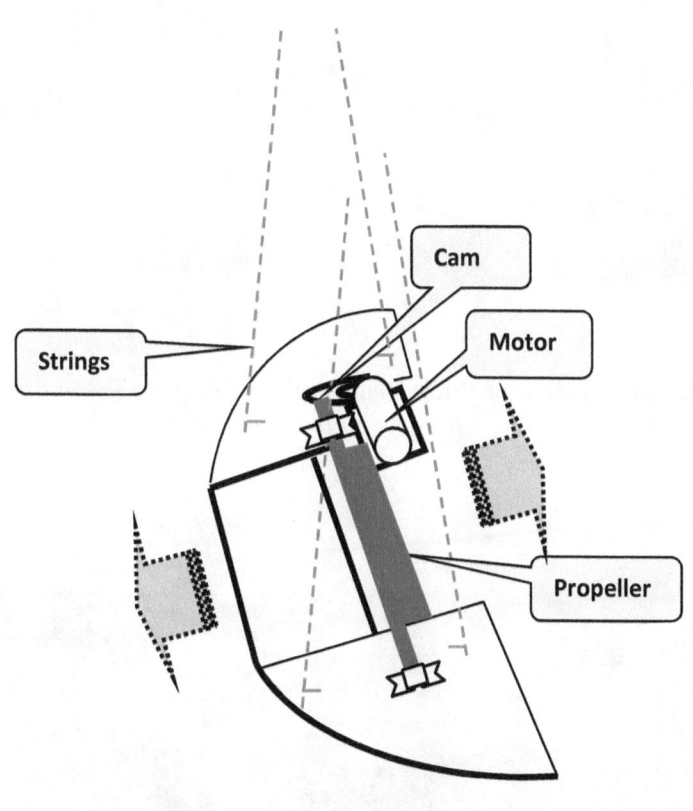

Schematic Diagram of Wave Propeller in Air

Demonstration of Wave Propeller in Air (*in the laboratory-library in my house*)

Wave Propeller in Air

Promising Applications

Low Noise Micro Size Aircraft

A low noise, lightweight, low power, slow flying, cheap micro size aircraft may be fashioned around the transverse wave propeller.

A conventional micro size aircraft is propelled by an ordinary propeller that is prone to high noise resulting from tip speed (that is proportional to both rpm and diameter). The transverse wave propeller may overcome the noise problem to an extent by imitating bird flight. The associated loss of energy from noise may also be conserved.

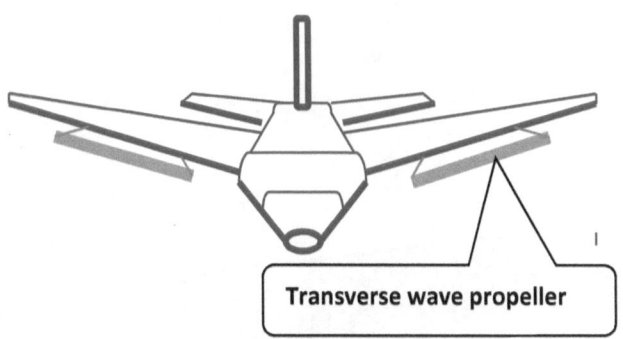

Transverse wave propeller

A Failed Experiment with Unpatriotic Fish

I dreamed of domesticating fish which may be encased in low head (say 6" head) 'live fish pumps'. Such 'live fish pumps' may be used in large numbers to pump water in paddy fields from plot to neighboring plot in stages. The fish may get a supply of nutrition from the water flowing through the pump casing. To demonstrate the idea of a live fish pump, I constructed a box with inlet and outlet openings and let water and fish in to the box:

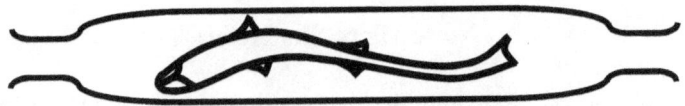

The fish refused to swim. They were unpatriotic. My dream of conserving power and diesel across India's vast countryside was dashed. (Not entirely yet; I still dream of succeeding in demonstrating the idea with the help of ichthyologists and geneticists who can help grow flat muscular bodied fish that quickly grow to adulthood and hold their size steady, pump water indefatigably and, after their useful life is over, provide gastronomic pleasure to the farmer and his family. Any volunteers?)

Building Wound-Core Transformer

How I Got This Idea

While undergoing training at BHEL, Bhopal, as a 4th year engineering student, I was sent to the transformer manufacturing section where very large transformers, some a few hundred MW, were being built. Looking at the core building process which involves a great deal of precision machining, I wondered why is it necessary to expend so much engineering effort on an object which is mechanically static, free of moving parts. There I got this idea. I applied for and got patent rights for the same. I also published an article on the subject in the well known journal, Electrical India. The text of the article is reproduced below:

Beginning of Text:

Building Wound-Core Transformer

by

Ligaraj Patnaik, M.I.E.

Abstract:

A method of building transformers with continuous core is illustrated. This wil have obvious theoretical and practical advantages. Time and cost of manufacturing will be reduced. The manufacturing can be completely machanised.

Concept:

Building the Core

PLEASE REFER Fig. 1. Buildig the core can be accomplished wih a mould having two straight sides. Avoiding sharp bends in the core will help minimise the internal stresses. The mould may be mounted on a horizontal turn table. Care must be exercised to achieve the two straight sides of the core as precisely as possible. Spring loaded idling rollers exerting lateral pressure may be used for the purpose. At the end of the operation tight bracings must be used to maintain the shape.

Conductor Winding

Please refer Fig. 2. For winding the conductors, a splittable mould will be introduced about one straight side of the core. It can be held floating with the help of various rollers. The mould may have, on one or both the circular ends, means to attach a belt, a chain or a gear to drive it. Then suitable conductor can be wound upon it. The mould wil be left behind in the transformer. When the occasion arises for repairing the transformer, the winding can be unwound with the help of this mould. By this means the other winding(s) of either one phase or three phase transformers can be built.

Building Wound-Core Transformer

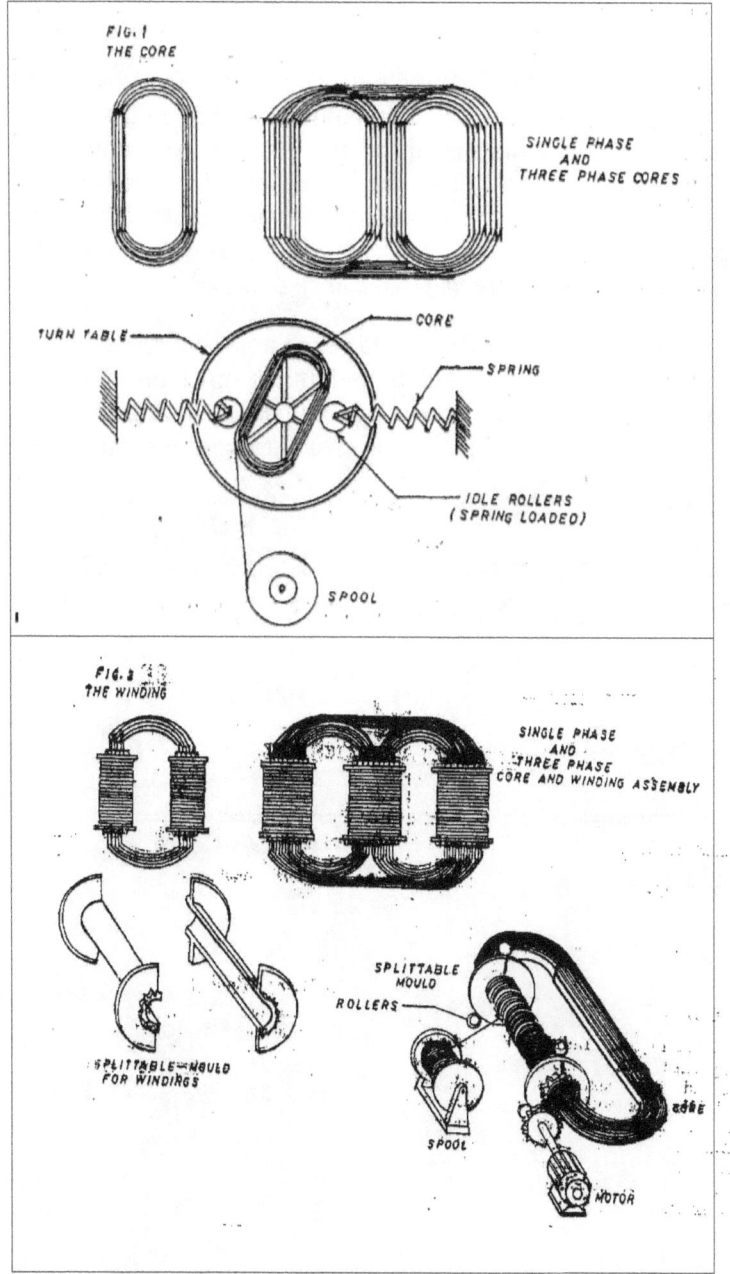

Advantages

Employing this method of construction will help avoid many of the cutting operations like cropping, piercing, mittring. Also this will help avoid much of the manual operations like unlacing, relacing, etc. This method of building can be more easily automated than the conventional one. Besides the above, real benefits like (i) continuous, hence, more efficient core and (ii) less wastage of core material can be obtained.

Leaving behind splittable moulds need not present any special problems. In fact, this will be helpful in unwiding damaged windings and rewinding new one without requiring unlacing and relacing, thus saving on time, effort and cost of repair work.

Resume:

Existing manufacturing lines can be considered for adoption of this method of core building.

Existing transformers, when beyond repair, can be replaced with new ones.

===

End of Text.

Solution to Nuisance Tripping of Plant Lightning from Fault in Appliances

How I Got This Idea

Fed up of me for insisting on resetting the earth leakage circuit breakers (ELCBs) every time these tripped, causing lights and fans to go off, a very good friend of mine threatened to throw me and my table out of the shared work space. That set me thinking for a solution. I felt that ceiling fans and lights need not be covered by 30 mA ELCBs as these are generally handled by professional electricians. Only the appliances plugged to the wall sockets expose the lay users to the hazard of electricity. It is sufficient to cover only these wall sockets by the ELCBs leaving the ceiling fans and lights alone. The reputed journal 'Electrical India' have kindly published an article which is reproduced below.

Beginning of Text:

======================================

Solution to Nuisance Tripping of Plant Lighting from Fault in Appliances

By
Lingaraj Patnaik MIE

Solution to Nuisance Tripping of Plant Lightning from Fault in Appliances

Nuisance tripping of plant lighting circuits protected by Earth Leakage Circuit Breakers is a nightmare for electrical maintenance people especially when a portion of the plant lighting trips for no better reason than an earth leakage or earth fault in, say, a plug socket outlet.

Whereas safety is achieved upto 100%, circuit availability is reduced below acceptable levels. And this, in many instances, leads to bypassing of the ELCBs by the maintenance people.

The trouble arises because the same ELCB which potects the luminaire circuits also protects the power socket outlets meant for appliances.

Whereas luminaires do not develop earth faults frequently the same is not true for appliances hooked to power socket outlets where frequent earth faults do take place.

The solution to the above problems could be in the following directions:

1) All plug sockets and appliances should be fed from independent MCB DBs to be de-

signated as 'Appliance MCB DBs.' Only luminaire circuits should be fed from 'Luminaire MCB DBs'. If this is done plant lighting shall not suffer nuisance tripping owing to appliances.

2) If however, the same MCB DB is to be used for both luminaires and appliances then 2 phases should be used for luminaires and the 3rd phase should be reserved for plug sockets and appliances only. Of course plug & sockets should be kept in boards separate from luminaire switch boards.

3) 100 mA/300 mA ELCB may be used for the 'luminaire MCB DBs' or the 'Luminaire phases' (2 out of 3 phases) and 30 mA ELCB for the 'Appliance MCB DBs' or the 'Appliance phase' (3rd phase).

This will ensure fire hazard protection (100/200mA) for the luminaire circuits and personal safety (30 mA) for appliances circuits, and, thus, we feel, will provide the optimum performance for the combined availability and safety aspects.
==

End of Text.

Dogma and the Baconian Filter for Separating the Scientific from Speculative Theories

A Peaceful Evening in My Household

My young daughter and her cousin started a discussion on the solar system one evening at dinner time. We were squatting on the floor, Indian style. On being asked whether the sun revolves around the earth or the earth moves around the sun, both of them, wise in their sixth grade, stated emphatically that the sun stands still and the earth rotates around its axis. Sneeringly I asked, whether all of us appeared to be rotating? Whether the house appeared to be rotating? Or the trees appeared to be rotating? Or were our heads rotating? Visual evidence was against them. Playfully I pointed out that in their third grade they had studied and written in the examinations that the sun rises in the east and sets in the west. So both the teachers and the text books of the third standard were wrong, or else, the teachers and the text books of the sixth standard are wrong. Their pride was pricked. By this time, having finished my dinner, I was sitting on the sofa. Peeved, the children flew at me from either side and tried to convince me, by physically pinning me down, saying that I did not know anything, and threatened to take me to a scientist uncle who shall explain me everything. Subsequently the uncle assured them that the earth was, indeed, rotating around its axis and the sun is still. Point is, without accepting additional physical principles, such as the principle of 'minimum total energy of a system', one cannot determine with certainty which particular version is correct. The simple principle of 'equal but opposite mutual relative velocities' would admit of both the versions as equally correct, leaving the matter to be decided by each observer for himself. However, today, we are conditioned to abandon direct sense evidence lightly in favor of authority.

Dogma and the Baconian Filter for Separating the Scientific from Speculative Theories

CENTURIES OF SCIENCE has not helped us shed our habit of dogma. When one group of gods fall, another takes over and carries on as if infallible. It is time to submit every theory to the Baconian filter, as exemplified below, before we pronounce a theory to be either a scientific or, merely, speculative theory.

Fundamental Laws of Physics

The statement of every fundamental law of physics is a statement of faith. Take, for example, the laws of Newton, Maxwell's equations, postulates of the special and the general relativities of Einstein, Schrödinger's equations, Heisenberg's uncertainty principle, etc.

Once a law can be derived from other fundamental laws, it is no more a fundamental law, now it is a derived law. For example, the ideal gas laws are derived laws (derived from Newton's laws of motion in statistical mechanics). In order to be considered complete, a treatment of the fundamental basis of any theory must always be accompanied with *a checklist of limitations*.

Example 1

List of Limitations (Baconian Filter) of the Newtonian Mechanics:

	Points of examination	**Observations**
1	Errors/ difficulties/ paradoxes carried forward from tributary theories	
2	Errors/ difficulties/ paradoxes of precedent theories solved	
3	New errors/ difficulties/ paradoxes created	1. The terms space, time, energy, action, reaction are not defined. 2. The first law is actually a definition of

		inertia, a property of matter, like any other property of matter, for example, volume, density, etc. Newton, as a matter of fact, speaks of inertia as a definition before he stated it as a law of nature in his *Principia*. 3. The first law, indeed, can be derived as a special case of the second law of Newton for an external force $F = 0$.
4	Postulates/ additional list of postulates: tacit ones, generalizations made on the original set of the postulates, that lead to various results not reachable by the first set of postulates	
5	Mathematical liberties taken/ limitations accepted	
6	Claims	
7	Disputes	
8	Experiments proposed for verification of the predicted results and the extent to which the theory shall stand verified	
9	Extent of experimental verification, repeatability and ranges of various errors	

10	Possible alternative theories/ suggestions/ recommendations	
11	Examination of theoretical contribution in terms of	
	(a) Simplification of method	
	(b) Simplification of presentation	
	(c) Revision of existing fundamentals	
	(d) Further development of existing theories	
	(e) Consolidation of different existing theories	
	(f) Mere new definitions, symbols	
	(g) Mere indulgence in speculations which cannot be or has not been verified	
12	Practical applications	

I have left gaps which the reader may attempt to fill.

Example 2

List of Limitations (Baconian Filter) of the Theory of Electromagnetism of Maxwell:

	Points of examination	**Observations**
1	Errors/ difficulties/ paradoxes carried forward from tributary theories	
2	Errors/ difficulties/ paradoxes of precedent theories solved	1. Electricity and magnetism were

		combined. 2.　$c = $ constant was derived for electromagnetic waves.
3	New errors/ difficulties/ paradoxes created	4.　Redundancy: There are two unknowns **E** and **B** but four equations: The two *divergence* equations can easily be derived from the two ***curl*** equations. 5.　The physical reality of the Poyinting vector is not established.
4	Postulates/ additional list of postulates: tacit ones, generalizations made on the original set of the postulates, that lead to various results not reachable by the first set of postulates	
5	Mathematical liberties taken/ limitations accepted	1.　Lack of mathematical methodology forcing the choice of separation of variables by way of the (Ludwig) Lorenz gauge.
6	Claims	
7	Disputes	1.　Lack of ready compatibility between Maxwell's equations and $E = h\nu$ for photons.
8	Experiments proposed for verification of the predicted	

	results and the extent to which the theory shall stand verified	
9	Extent of experimental verification, repeatability and ranges of various errors	
10	Possible alternative theories/ suggestions/ recommendations	
11	Examination of theoretical contribution in terms of	
	(a) Simplification of method	
	(b) Simplification of presentation	
	(c) Revision of existing fundamentals	
	(d) Further development of existing theories	
	(e) Consolidation of different existing theories	
	(f) Mere new definitions, symbols	
	(g) Mere indulgence in speculations which cannot be or has not been verified	
12	Practical applications	Endless.

Example 3

List of Limitations (Baconian Filter) of the Theory of Special Relativity:

	Points of examination	**Observations**
1	Errors/ difficulties/ paradoxes	

	carried forward from tributary theories	
2	Errors/ difficulties/ paradoxes of precedent theories solved	1. Explained constancy of the velocity of light from the Michelson- Morley experiment.
3	New errors/ difficulties/ paradoxes created	1. Either the special theory of relativity, which asserts that no finite rest mass can move at the speed of light, is correct or else, the traditional rule of 'equal but opposite mutual relative velocities between any two objects' is correct because I routinely move at the velocity of light past every photon streaming in my neighborhood.
4	Postulates/ additional list of postulates: tacit ones, generalizations made on the original set of the postulates, that lead to various results not reachable by the first set of postulates	1. c = constant. 2. The generalization of $dr^2 - c^2 dt^2 = 0$ to $dr^2 - c^2 dt^2 = ds^2$ where ds is defined as a space-time interval and postulated as an invariant is a stroke of intuition of Einstein. This may be seen as the central postulate of the special theory of relativity from which the facts (i) c = constant and (ii) the Special (Hendrik) Lorentz Transformations

		can be derived. Bypassing this postulate, the Special (Hendrik) Lorentz Transformations cannot be derived from c = constant. Thus, $dr^2 - c^2 dt^2 = ds^2$, with the advantage of hindsight, may be seen as the central postulate of the special theory of relativity.
5	Mathematical liberties taken/ limitations accepted	
6	Claims	
7	Disputes	
8	Experiments proposed for verification of the predicted results and the extent to which the theory shall stand verified	
9	Extent of experimental verification, repeatability and ranges of various errors	
10	Possible alternative theories/ suggestions/recommendations	
11	Examination of theoretical contribution in terms of	
	(h) Simplification of method	
	(i) Simplification of presentation	
	(j) Revision of existing fundamentals	
	(k) Further development of existing theories	

	(l) Consolidation of different existing theories	
	(m) Mere new definitions, symbols	
	(n) Mere indulgence in speculations which cannot be or has not been verified	
12	Practical applications	

A Dream By Way Of Conclusion

I dream of a large world library for fundamental theories, built with a tunnel like architecture, where theories dealing only with the foundations of physics find place. Only those theories that pass the Baconian filter will find entrance in to the sanctum sanctorum of 'scientific' theories.

No doubt, refinements shall be made to the Baconian filter from time to time. No theory will stand the scrutiny of one or all the versions of Baconian filters always. Consequently all the theories are apt to be classified as part scientific and part speculative at any given moment of time. Besides, the candidate theories may be shuffled back and forth between the two chambers from time to time and this shall represent the vigor of the current scientific activity. We must never forget that where as a collection of facts of nature comprises a discovery, a theory is always an invention of the human mind. No theory, hence, shall attain the status of a perfect scientific theory forever.

The Dream Library:

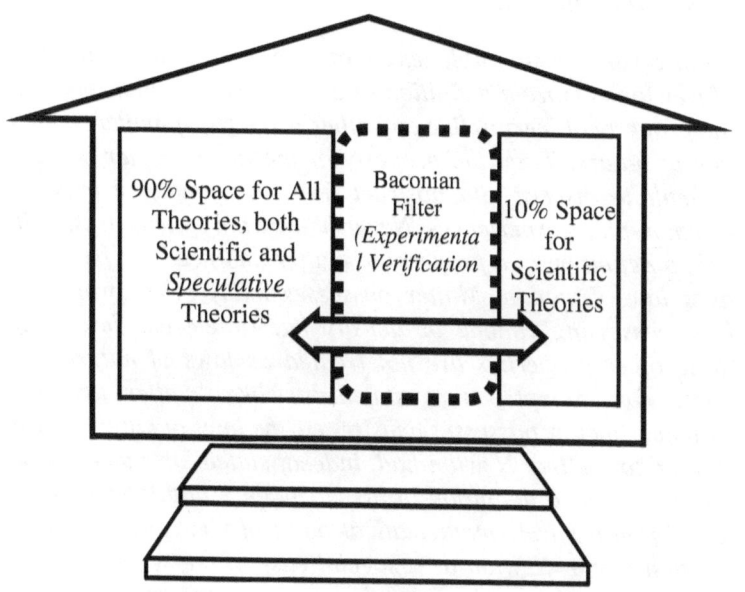

Derivation of the Three Laws of Newton as Corollaries of the Law of Conservation of Energy

How I Got This Idea

For a very considerable length of time, I felt that the first law of Newton is more a definition of a property of matter, namely, the property of inertia, like any other property of matter, than a law of nature. True, the property of inertia of matter is quite difficult to discern and abstract from the confusing mass of common place experiences. Nevertheless, the statement is, still, only a definition of a property of matter and need not be stated as a law of motion. Matter possesses many other properties like, possessing volume or density but statements describing these other properties are not treated as laws of nature. We don't, for example, treat statements like 'matter occupies volume', 'matter possesses density' etc. as laws of nature. I was thrilled to see that Newton had, indeed, stated his statement on inertia first as a definition in his 'Principia' and later on as a law. From a purely mathematical point of view one may treat the first law of Newton as a special case of the second law by setting force 'F' equal to zero for a given mass 'm', and, thus, obtain zero acceleration 'a' which implies uniform velocity, the first law of motion.

ONE TIME, I noticed that writing $(-)dV/dr$ for force F in the 2nd law of Newton, where 'V' is potential energy, makes the statement appear as a statement of conservation of energy (mechanical energies): $F = ma$; $(-)dV/dr = ma = mdv/dt$; $(-)dV = mdv (dr/dt) \equiv mvdv \equiv md(\frac{1}{2}v^2)$ i.e., $(-)dV = d(\frac{1}{2}mv^2)$; $d(\frac{1}{2}mv^2) + dV = 0$; writing $\frac{1}{2}mv^2$ as 'T', kinetic energy, we receive $d(T + V) = 0$ i.e. $T + V$ is a constant, which is the statement of the law of conservation of energy for mechanical energies. I toyed around and derived all the three laws of Newton from the law of conservation of

Derivation of the Three Laws of Newton from the Law of Conservation of Energy

energy for the special case of mechanical energies (potential + kinetic). Here follow the derivations:

Kinetic and Potential Energies

We start with a force field complete with various bodies in it at rest or in motion. We associate a form of energy with the movement of bodies and choose the term kinetic energy, T, to represent this energy as a function of speed:

$$\text{Kinetic energy, } T \equiv e_1(v) \qquad (1)$$

We also associate another form of energy with the force field and choose the term potential energy, V, to represent this energy as a function of position:

$$\text{Potential energy, } V \equiv e_2(s) \qquad (2)$$

Starting With the Law of Conservation of Energy

We start with assuming that the law of conservation of energy is true: in an isolated ensemble of objects in a field (comprising a system) the total energy remains constant indefinitely:

$$T + V = \text{constant} \qquad (3)$$

Deriving the First Law of Newton

For an isolated body (i.e., in the absence of force field),

$$V = 0.$$

Thus, for an isolated body, in the absence of a force field, the law of conservation of energy (3) boils down to

$$T = \text{constant} \qquad (4)$$

But kinetic energy is $T = e_1(v)$ from definition (1).

Derivation of the Three Laws of Newton from the Law of Conservation of Energy

Thus the function $e_1(v)$ is constant. This implies velocity v is constant for an isolated body in the absence of an external force field.

This is the first law of Newton.

Alternatively, we may, derive the second law $F = ma$ first and then consider the special case of $F = 0$ to get $a = 0$ implying uniform velocity, for the first law of Newton.

Deriving the Second Law of Newton

Starting with the law of conservation of energy,

$$T + V = \text{constant} \qquad (3)$$

i.e., $d(T + V) = 0$

i.e., $d(\tfrac{1}{2} mv^2) = (-)d(V)$

i.e., $mv\, dv = (-)\, d(V)$

Now, we know, force $f = (-)\, d(V)/dr$ *(the space rate of decrease of potential energy)* (5)

Thus, $mv\, dv/dr = (-)\, d(V)/dr = f$

Thus, $f = mv\, dv/dr$

Multiplying RHS by dt/dt and simplifying, we receive

$f = mv\,(dv/dr)(dt/dt) \equiv mv\,(dv/dt)(dt/dr) \equiv mv(a)(1/v) \equiv ma$

i.e., $f = ma$ \qquad (6)

This is the second law of Newton.

Derivation of the Three Laws of Newton from the Law of Conservation of Energy

Deriving the Third Law of Newton

Coming back to the law of conservation of energy for an isolated system,

$$T + V = \text{constant} \qquad \text{from (3)}$$

Within the isolated system consider an exchange of energy de from the field to a body. An increment of the T of the body is accompanied with a decrement of the V of the system. Call the force, felt by the body and effecting the increment of T the 'active force':

$$f(\text{active})\, ds = d(T) = de \qquad (7)$$

Call the force, felt by the energy field (as a reaction from the body) mediating the decrement of V, the "reactive" force;

$$f(\text{reactive})\, ds = d(V) = (-)de \qquad (8)$$

From (7) & (8) we receive

$$f(\text{active}) + f(\text{reactive}) = d(T + V) = 0$$

which implies

$$f(\text{active}) = (-)f(\text{reactive}) \qquad (9)$$

This is the third law of Newton.

Seductive Nature and Her Laws

Nature is seductive. Laws reveal her secrets.

A Pressure Model for the Inverse Square Law

How I Got This Idea

When we compare electricity with fluid dynamics we notice a peculiar asymmetry in the terminology. In fluid dynamics, pressure is the motive and flow is the movement. In electricity, however, both the flux (motive) and the current (movement) are flow terms. This unfortunate situation has come about because the flow model has provided us with the inverse square law. Consider a gas, issuing out of a point source and expanding spherically in space. The rate of flow of the gas crossing a unit area of the spherical surface shall decrease inversely as the square of the radius. I felt compelled to look for a pressure model as an alternative while I was walking past my hostel at IIT-BHU one evening.

Flux

THE WORD *flux* comes from Latin: *fluxus* means "flow". In the field of electromagnetism, flux is the integral of a vector quantity, flux density, over a finite surface.

Maxwell says, "*In the case of fluxes, we have to take the integral of the flux, over a surface, through every element of the surface. The result of this operation is called the surface integral of the flux. It represents the quantity which passes through the surface*".

Traditional Derivation of the Inverse Square Law

If we consider a spherically symmetrical incompressible streamline fluid flow in a three dimensional space, far away from its 'point' source, we may write, for radii R and R',

∫flux density d(surface) at R = ∫flux density d(surface) at R'

i.e., (flux density at R)*4 π R² = (flux density at R')*4 π R'²

i.e., (flux density at R)/(flux density at R') = 4 π R'² /4 π R²

i.e., flux density ∝ 1/R²

A Pressure Model for the Inverse Square Law

Consider an elastic solid, sliced along radii from the point source of pressure so that pieces can slide past each other without generating shear forces. Consider a free body diagram for the body A shown below. As the body A is not accelerating, the forces shown (as arrows) are same in magnitude but are opposite in direction The areas are γr_1^2 and γr_2^2 respectively for the portion of spheres with in the solid angle γ (or, $4\pi r_1^2$ and $4\pi r_2^2$ respectively for the whole spheres). Clearly the pressures are $F/\gamma r_1^2$ and $F/\gamma r_2^2$ respectively (inverse square law); i.e., the pressures decrease as per the inverse square law as the distance r increases, i.e., pressure $\propto 1/R^2$:

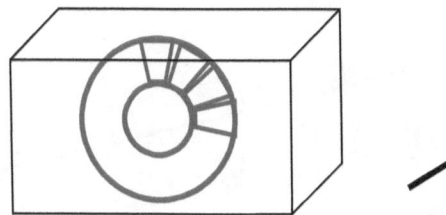 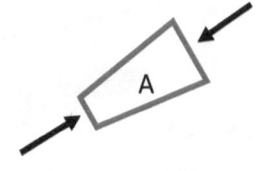

Conclusion

The pressure model is certainly a more sensible picture for the inverse square law than the flow model associated with the term 'flux'. However, may someone help find a suitable term to replace 'flux' in the current literature of electricity: will the good old term 'stress' be more suitable?

Reduction of the Set of Four Electromagnetic Equations of Maxwell to a Set of Two Equations

How I Got This Idea

*At Nagercoil, suddenly one day I asked myself why we require four equations of Maxwell to deal with only two fields, the electric and the magnetic. I felt, for mathematical conciseness, two equations should be sufficient. Preparing for a long struggle, it was a pleasant surprise for me to be able to derive the two divergence equations from the two **curl** equations in virtually no time.*

Introduction

MAXWELL'S EQUATIONS are a set of four partial differential equations that describe how the electric and magnetic fields relate to their sources, charge density and current density. Conventional symbols are used in this text:

div $\mathbf{E} = \rho/\varepsilon_0$ (1)

div $\mathbf{B} = 0$ (2)

curl $\mathbf{E} = -\partial_t \mathbf{B}$ (3)

c^2 **curl** $\mathbf{B} = \partial_t \mathbf{E} + \mathbf{j}/\varepsilon_0$ (4)

Indeed, we see that it is possible to derive the two *divergence* equations from the remaining two ***curl*** equations as follows:

Reduction of the Set of Four Equations to a Set of Two

Taking divergence of both the sides of the fourth equation we receive:

c^2 div **curl B** = ∂_t div **E** + div **j**$/\varepsilon_0$

But   div **curl B** is identically zero.

Thus $0 = \partial_t$ div **E** + div **j**$/\varepsilon_0$ i.e.,

∂_t div ε_0 **E** = - div **j**

But,   div **j** = - dρ/dt . Substituting in the last equation we immediately receive the first equation of Maxwell:

div **E** = ρ/ε_0 (1st equation)

Thus, we have derived the first equation of Maxwell from the fourth.

Now we shall, similarly, derive the second equation of Maxwell from the third. Taking the divergence of both the sides of the third equation we receive:

div **curl E** = - ∂_t (div**B**)

As before div **curl E** is identically zero. Thus

$0 = \partial_t$ div **B**   i.e.,

div **B** = a constant.

However, considering the fact that there are no magnetic monopoles, we set the constant to zero. Thus, we receive the second equation of Maxwell:

div **B** = 0   (2nd equation)

Thus, we have derived the second equation of Maxwell from the third.

Reduction of the Set of Four Electromagnetic Equations of Maxwell to a Set of Two Equations

Conclusion

We have, thus, reduced the set of four electromagnetic equations of Maxwell to a set of two electromagnetic equations:

<u>The final set of two electromagnetic equations:</u>

$$\text{curl } \mathbf{E} = - \partial_t \mathbf{B} \quad \text{and}$$
$$c^2 \text{ curl } \mathbf{B} = \partial_t \mathbf{E} + \mathbf{j}/\varepsilon_0$$

After making the above derivations, I learned that the same results had earlier been obtained by others. While many appreciated my work, a young friend of mine from BITS, Pilani showed a book from the internet where the same fact was mentioned. I felt somewhat letdown. However, the book said, some people consider the two divergence equations of Maxwell to provide boundary conditions for the two **curl** equations. I felt, in any case, the boundary conditions must be equations that are independent of the system equations. As the two divergence equations are derivable from the two **curl** equations, these are dependent and as such, cannot serve as boundary conditions.

SECOND PAPER:

Reduction of the System of Electromagnetic Equations of Maxwell in Tensor and Matrix Forms

*We have already reduced the system of four electromagnetic equations of Maxwell in vector form, involving **E** (electric field) and **B** (magnetic field) to a fundamental system of two equations in the earlier paper. We shall, now, reduce the familiar system of eight electromagnetic equations of Maxwell in partial differential form to a fundamental system of six equations in six variables, E_x, E_y, E_z, B_x, B_y, B_z. We shall also write the same system in tensor and matrix forms.*

Reduction of the System of Maxwell's Equations in Partial Differential Form

EXPRESSING MAXWELL'S EQUATIONS in partial differential form, we receive the following system of eight equations in six variables E_x, E_y, E_z, B_x, B_y, B_z:

$\partial_x E_x + \partial_y E_y + \partial_z E_z = \rho/\varepsilon_0$ (1a)

$\partial_x B_x + \partial_y B_y + \partial_z B_z = 0$ (2a)

$\partial_y E_z - \partial_z E_y = -\partial_t B_x$ (3a)

$\partial_z E_x - \partial_x E_z = -\partial_t B_y$ (3b)

$\partial_x E_y - \partial_y E_x = -\partial_t B_z$ (3c)

$\partial_y B_z - \partial_z B_y = (\partial_t E_x + I_x/\varepsilon_0)c^2$ (4a)

$\partial_z B_x - \partial_x B_z = (\partial_t E_y + I_y/\varepsilon_0)c^2$ (4b)

$\partial_x B_y - \partial_y B_x = (\partial_t E_z + I_z/\varepsilon_0)c^2$ (4c)

SECOND PAPER: Reduction of the System of Electromagnetic Equations of Maxwell in Tensor and Matrix Forms

As the (divergence) equation (1a) may be derived by taking further partial derivatives of the (**curl**) equations (3a), (3b) and (3c) and adding up, and similarly, the (divergence) equation (2a) may be derived from the (**curl**) equations (4a), (4b) and (4c), we can reduce the above to a system of six equations consisting of only the last six equations.

Thus, ignoring the (divergence) equations (1a) and (2a) above, and collecting the rest six equations together, we have reduced the familiar system of eight partial differential electromagnetic equations of Maxwell to a fundamental system of six partial differential equations in six variables E_x, E_y, E_z, B_x, B_y, B_z.

Maxwell's Equations in Tensor Form

In familiar tensor form the Maxwell's equations are written as follows [*equations* (5) *and* (6)]:

$$\partial_\nu \psi_{\mu\nu} = I_\mu \qquad (5)$$

[*This equation is Maxwell's equations* 1(a), (4a), (4b), (4c) *in partial differential form taken together; this tensor equation* (5) *contains the two vector equations* (1) *and* (4) *of the earlier paper (vector equation* (1) *is redundant and was derived from vector equation* (4) *in the earlier paper)*]

$$\partial_\sigma \psi_{\mu\nu} + \partial_\mu \psi_{\nu\sigma} + \partial_\nu \psi_{\sigma\mu} = 0 \qquad (6)$$

[*This equation is Maxwell's equations* 2(a), (3a), (3b), (3c) *in partial differential form taken together, this tensor equation* (6) *contains the two vector equations* (2) *and* (3) *of the earlier paper (vector equation* (2) *of the earlier paper is redundant and was derived from vector equation* (3) *of the earlier paper)*]

The tensor equation (5) can be written in matrix form as follows [*this, immediately, yields the equations* (1), (4a), (4b), (4c)]:

SECOND PAPER: Reduction of the System of Electromagnetic Equations of Maxwell in Tensor and Matrix Forms

Maxwell's Equations in Matrix Form

$$\begin{pmatrix} 0 & B_z & -B_y & -iE_x/c \\ -B_z & 0 & B_x & -iE_y/c \\ B_y & -B_x & 0 & -iE_z/c \\ iE_x/c & iE_y/c & iE_z/c & 0 \end{pmatrix} \begin{pmatrix} \partial_x \\ \partial_y \\ \partial_z \\ \partial_{ict} \end{pmatrix} = \begin{pmatrix} I_x/\varepsilon_0 c^2 \\ I_y/\varepsilon_0 c^2 \\ I_z/\varepsilon_0 c^2 \\ i\rho/\varepsilon_0 c \end{pmatrix} \equiv I_\mu \quad (5a)$$

Similarly, the tensor equation (6) can be written in matrix form as follows [*it is a little strange looking considering the sequence of* ∂_x, ∂_y, ∂_z, *and* ∂_{ict} *but it yields the equations* (2), (3a), (3b), (3c) *correctly*]:

$$\begin{pmatrix} 0 & B_x & B_y & B_z \\ -B_x & 0 & iE_z/c & -iE_y/c \\ -B_y & -iE_z/c & 0 & iE_x/c \\ -B_z & iE_y/c & -iE_x/c & 0 \end{pmatrix} \begin{pmatrix} \partial_{ict} \\ \partial_x \\ \partial_y \\ \partial_z \end{pmatrix} = \begin{pmatrix} 0 \\ 0 \\ 0 \\ 0 \end{pmatrix} \quad (6a)$$

where the electromagnetic tensor is written as appearing in the equation (5a).

SECOND PAPER: Reduction of the System of Electromagnetic Equations of Maxwell in Tensor and Matrix Forms

$$\psi_{\mu\nu} \equiv \begin{pmatrix} 0 & B_z & -B_y & -E_x/c \\ -B_z & 0 & B_x & -E_y/c \\ B_y & -B_x & 0 & -E_z/c \\ E_x/c & E_y/c & E_z/c & 0 \end{pmatrix} \quad (7)$$

However, each of the tensor or matrix equations in four dimensions shall yield four equations each, together yielding eight equations, and are, together, not suited to express the fundamental system of six partial differential equations in six variables $E_x, E_y, E_z, B_x, B_y, B_z$ free of redundancy.

The tensor equations in three dimensions are eminently suitable to represent the fundamental system of six equations in six variables $E_x, E_y, E_z, B_x, B_y, B_z$:

$$\begin{pmatrix} 0 & B_z & -B_y \\ -B_z & 0 & B_x \\ B_y & -B_x & 0 \end{pmatrix} \begin{pmatrix} \partial_x \\ \partial_y \\ \partial_z \end{pmatrix} = (1/c^2) \begin{pmatrix} \partial_t E_x + I_x/\varepsilon_0 \\ \partial_t E_y + I_y/\varepsilon_0 \\ \partial_t E_z + I_z/\varepsilon_0 \end{pmatrix} \quad (5b)$$

$$\begin{pmatrix} 0 & iE_z/c & -iE_y/c \\ -iE_z/c & 0 & iE_x/c \\ iE_y/c & -iE_x/c & 0 \end{pmatrix} \begin{pmatrix} \partial_x \\ \partial_y \\ \partial_z \end{pmatrix} = \begin{pmatrix} -\partial_t B_x \\ -\partial_t B_y \\ -\partial_t B_z \end{pmatrix} \quad (6b)$$

SECOND PAPER: Reduction of the System of Electromagnetic Equations of Maxwell in Tensor and Matrix Forms

Conclusion

In the earlier (first) paper we had reduced the system of four electromagnetic equations of Maxwell in vector form to a fundamental system of two equations. In this paper we have reduced the familiar system of eight electromagnetic equations of Maxwell in partial differential form to a fundamental system of six equations in six variables. We also found that in tensor and matrix forms in four dimensions each tensor equation necessarily contains one redundant partial differential equation (which are the divergence equations in vector form) whereas in three dimensions both the tensor and matrix forms satisfactorily represent the fundamental system of six partial differential equations in six variables, E_x, E_y, E_z, B_x, B_y, B_z free of redundancy. However, the tensor $\psi_{\mu\nu}$ defined in (7) is in keeping with this current fashion but we must keep the fact of redundancy in the back of our mind.

The Questionable Freedom of Fixing the (Ludwig) Lorenz Gauge

How I Got This Idea

For quite some time I was not comfortable with the Maxwell's equations. The "fortunate" freedom of choice of fixing the (Ludwig) Lorenz gauge was a reason. The (Ludwig) Lorenz gauge is a device to effect a separation of the two variables $\mathbf{A}$ and ϕ (related to $\mathbf{E}$ and $\mathbf{B}$; look at the defining equations (4) and (4') below), a method for solving the Maxwell's equations somehow in the absence of a better mathematical method. There is no freedom of choice here.

ALL THE SYMBOLS in this essay have currently familiar meanings.

Whereas two equations should be sufficient for relating the two variables $\mathbf{E}$, the electric field, and $\mathbf{B}$, the magnetic field, four equations are available in the form of Maxwell's equations (historically, Oliver Heaviside gave us the vector form of the four electromagnetic equations) and we have the problem of redundancy. As a matter of fact the two divergence equations are easily derived from the two **curl** equations (by taking divergences). The two **curl** equations provide the mathematically minimum basis for Maxwell's equations:

curl $\mathbf{E}$ = (-)$\partial \mathbf{B}/\partial t$ (1)

curl $c^2\mathbf{B}$ = $\partial \mathbf{E}/\partial t$ + $\mathbf{j}/\varepsilon_0$ (2)

div $\mathbf{B}$ = 0   (from divergence of (1))   (1')

div $\mathbf{E}$ = ρ/ε_0 (from divergence of (2))   (2')

(This set of four equations is redundant and this is perhaps the reason why authors of various books on higher mathematics avoid discussing this set of equations).

Familiar Steps Leading to the (Ludwig) Lorenz Gauge; an Admission of Lack of Adequate Mathematical Methods

Starting with (1'), as the divergence of **curl** is zero, we may define

B ≡ **curl A**   *(definition)*   (4)

where **A** is called the vector potential.

Substituting this *definition* of **B** in equation (1) we receive,

curl E = (-)∂(**curl A**)/∂t . This allows us to define

E ≡ (-)∂**A**/∂t + (-)∇ϕ *(definition)*   (4')

taking advantage of the fact that *curl* of *gradient* is zero. Now we may substitute **B** and **E** from equations (4) and (4') in to (2) and receive

c^2 **curl curl A** = ∂[(-)∂**A**/∂t + (-)**grad** ϕ]/∂t + **j**/ε_0

Making use of the following familiar identity **curl curl A** ≡ **grad**(div **A**) + (-)∇²**A** we receive

c^2[**grad**(div **A**) + (-)∇²**A**] = (-)∂²**A**/∂t² + (-)∂(**grad** ϕ)/∂t + **j**/ε_0

Collecting the gradient terms in the RHS, we receive,

(-)c^2∇²**A** + ∂²**A**/∂t² = (-) c^2∇(div **A**) + (-) ∂(**grad** ϕ)/∂t + **j**/ε_0

i.e.,

(-)c^2∇²**A** + ∂²**A**/∂t² = (-)**grad** [c^2div **A** + ∂ϕ/∂t] + **j**/ε_0 (5)

This term [c^2div **A** + ∂ϕ/∂t] is the crux of the matter.

The (Ludwig) Lorenz gauge consists in setting this term, the expression with in the square brackets, to zero, i.e.,

$[c^2 \text{div } \mathbf{A} + \partial\phi/\partial t] = 0$ *(Ludwig Lorentz gauge)* (6)

This measure helps to simplify the matters mathematically, and helps to decouple $\mathbf{A}$ and ϕ in the complicated equation (5). The decoupled equations are

$$\nabla^2 \mathbf{A} - (1/c^2)\partial^2 \mathbf{A}/\partial t^2 = (-)\mathbf{j}/\varepsilon_0 c^2 \quad (7) \quad \text{and}$$

$$\nabla^2 \phi - (1/c^2) \partial^2\phi/\partial t^2 = (-)\rho/\varepsilon_0 \quad (7')$$

Look at the definitions (4) and (4') which we reproduce below:

$\mathbf{B} \equiv \text{curl } \mathbf{A}$ (definition) (4) and

$\mathbf{E} \equiv (-)\partial \mathbf{A}/\partial t + (-)\text{grad } \phi$ (definition) (4')

(in charge-free space **grad** ϕ is zero).

Whereas vector "potential" $\mathbf{A}$ is a non-conservative field, scalar ϕ is a conservative field.

(Ludwig) Lorenz gauge is a compromise in the circumstances, *an admission of lack of adequate mathematical methods to solve the equation* (5). It is not a demonstration of any freedom of choice. There is no other choice available today which would help solve the equation (5) completely satisfactorily. The measure of the (Ludwig) Lorenz gauge results in two wave equations (7) and (7') above involving φ and $\mathbf{A}$.

A Cosmetic Conclusion: Improving the Appearance of Maxwell's Equations

The appearance of Maxwell's equations can be made a little more neat (i.e., by suppressing the following three mathematical objects: the minus sign, ε_0, and c) by adopting the following definitions:

The Questionable Freedom of Fixing the (Ludwig) Lorenz Gauge

$$\left.\begin{array}{l}\mathcal{E} \triangleq \varepsilon_0 \mathbf{E} \\ \mathcal{B} \triangleq ic\varepsilon_0 \mathbf{B} \\ \tau \triangleq (-)ict \, ; \, i \equiv \sqrt{(-1)} \\ \jmath \triangleq i\mathbf{j}/c \equiv \rho \, i\mathbf{v}/c \end{array}\right\} \text{(definitions)}$$

With the above definitions, Maxwell's equations appear (free from the minus sign, ε_0, and c) as:

$$\text{curl } \mathcal{E} = \partial \mathcal{B}/\partial \tau \tag{1}$$

$$\text{curl } \mathcal{B} = \partial \mathcal{E}/\partial \tau + \jmath \tag{2}$$

$$\text{div } \mathcal{B} = 0 \quad \textit{(redundant)} \tag{1'}$$

$$\text{div } \mathcal{E} = \rho \quad \textit{(redundant)} \tag{2'}$$

The new look is simpler compared to the familiar look of the Maxwell's equations. However, this new look is merely of a cosmetic value and adds little to the physical content of the theory.

A Conceptual Error in the Calculation of Length Contraction in Special Relativity

Core Idea

By the time you are reading this line, its image is already 2 nanoseconds old. When I notice a friend eight feet away he is already eight nanoseconds older. When I hold a foot scale in my hand and see it edgewise I not only see a span of space one foot long, but I also 'see' an associated span of time equal to a nanosecond. Indeed I suffer an illusion when I think that I see the entire foot scale as it exists at a single moment of time. The image of the far end of the foot scale reaching my retina is necessarily a nanosecond older than the image of the near end of the foot scale reaching my retina. This is a fact of physics. To stretch the point a little farther, it is indeed true to say that, in the event the far end of the foot scale explodes, I will notice the event only after a nanosecond later. However, for the duration of a nanosecond, I will be thinking that the foot scale is intact. This is because the image of the far end seen at the moment is a historical image. (If it is difficult to associate nanosecond responses with physiologically slow eyes, think in terms of extremely fast light sensing devices). My perception of a span of space is always, necessarily, accompanied with a historical span of time. This is the first of the two influences of time on our perception of space. The second influence of time on our perception of space arises from bodies moving at great relative speeds comparable to the speed of light.

We shall explore the core idea of this chapter after a small detour.

For a very long time I have wondered about my relative speed with respect to the photons streaming past me at 300,000 km per second in my neighborhood. I cannot see how it can be anything different from the same 300,000 km per second but in the opposite direction. After all, if P goes at a speed v with respect to Q, then Q must go at the same speed v with

A Conceptual Error in the Calculation of Length Contraction In Special Relativity

respect to P but in the opposite direction. Clearly either this principle of 'mutually same relative speed between any two observers but in opposite directions' is wrong or else special relativity is wrong. Both are not compatible with each other. After all the photons are physical objects, packets of energy, moving about in the physical space in my neighborhood. After this detour, let us go back to the core idea of this chapter.

First we shall set forth the core idea of this chapter in full. Later we shall elaborate the details at leisure. Refer the figure with two complex frames placed at an angle ψ between them.

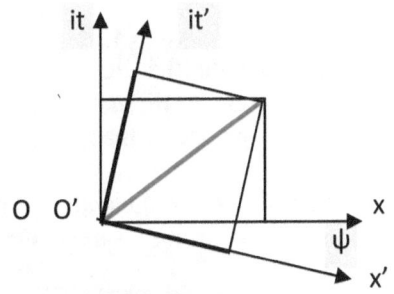

Complex space-time (x, it) and (x', it')

The primed frame (x', it') is the proper frame of a rigid rod and a clock. The unprimed frame (x, it) is the proper frame of an observer. The velocity of light is c = 1 unit. We write $\gamma \equiv \sqrt{[1 - v^2]}$ where v is expressed in units of c.

The familiar (Hendrik) Lorentz transformation for length is: $\Delta x = \gamma(\Delta x' + v\Delta t')$ i.e., $\Delta L = \gamma(\Delta L' + v\Delta T')$.

The line of argument in calculating the length contraction is this: at definite time t = 0 in the unprimed frame, the position of the integer x' = 1, with respect to the unprimed frame, is given by $x = \sqrt{(1 - v^2)} < 1$ implying contraction of length. This

A Conceptual Error in the Calculation of Length Contraction In Special Relativity

is popularly expressed as $L' = \gamma L$ meaning $L < L'$ as γ is always greater than one.

This line of argument is flawed. Let us analyze the matter in light of (a) the familiar transformation of length mentioned a while ago: $\Delta L = \gamma(\Delta L' + v\Delta T')$ and (b) the physical fact that my perception of a span of space is always, necessarily, accompanied with a historical span of time.

The integer $x' = 1$ is the point $(x', it') = (1, i0)$ in the primed frame.

The position $x = \sqrt{(1 - v^2)}$ at time $t = 0$ is the point $(x, it) = (1, i0)$ in the unprimed frame.

Both these points are outside the light triangles in their respective frames. By themselves, these two points represent nothing meaningful in the physical space-time. In order to represent a physically meaningful interval in the physical space, a mathematical projection on x-axis must always be associated with a mathematical projection on the it-axis such that $|\Delta t| \geq |\Delta x|$. Remember, $c = 1$.

Indeed, both these points in the respective complex spaces are in the region, outside the light triangles, which is named as 'elsewhere'. 'Nowhere-never' would also be a correct description. Every point inside the light triangles shall satisfy the condition $|\Delta t'| \geq |\Delta x'|$ or $|\Delta t| \geq |\Delta x|$ in order to meaningfully correspond to, say, a rigid rod, its span of length complete with its associated span of time, in the physical space-time.

Yes, $x' = 1$ is a mathematical projection on the x'-axis but, it is NOT a physical entity. A mathematical projection can be outside the light triangle, but a physical entity is always within the light triangle, inalienable from its associated mathematical projection along the imaginary time axis. For the rigid rod the projection on the time axis must be $|\Delta t'| \geq |\Delta x'|$ i.e., $|\Delta t'| \geq 1$. Thus, for the observer in the unprimed frame, the mathematical projection, $x = \sqrt{(1 - v^2)}$, represents contraction of only the

A Conceptual Error in the Calculation of Length Contraction In Special Relativity

mathematical projection $x' = 1$ and NOT that of a physical entity, like a rigid rod, complete with its projection of the associated span of time, $t' = 1$.

The correct equation for Δx must, at once, involve both the $\Delta x'$ and $\Delta t'$ components, i.e.,

$$\Delta x = [\Delta x' + v\Delta t']/\sqrt{[1 - v^2]} \text{ i.e.,}$$

$L = \gamma (L' + vT')$. We must recognize that T' cannot be dissociated from the rigid rod in the primed frame.

This is the core idea of this chapter. The rest is an elaboration and finding some consequences.

Physical Space

A mathematical space is useful, for example, to make an idealized but an approximate model of the physical space. The physical space exhibits certain unique features unlike a mathematical space. Fixed laws of nature operate in the physical space. Time passes in it. Gravitational and other fields manifest in it. These various fields possess energy in the physical space. The physical space harbors matter. Electromagnetic waves pass through it at the fixed speed of 300,000 km per second. All these features are natural to the physical space. But these features are optional to a mathematical space.

Mathematical Spaces

One is free to propose arbitrary rules of game in a mathematical space (whereas the rules of the game in the physical space are fixed and are the objects of scientific discovery). Passage of time is not represented in a mathematical space until we attach a time-axis to it, at least implicitly. Fields, energy, waves or matter are added symbolically as the need may arise. One may freely speculate that for a given mathematical space $F = ma^2$ (whereas the rule $F = ma$ is the fixed rule that governs dynamics in the physical space).

A Conceptual Error in the Calculation of Length Contraction In Special Relativity

Classical Space: Mathematical Model of the Physical Space as Conceived in Classical Physics

Space in classical physics is conceived as a three dimensional entity. A universal time flows in it uniformly everywhere. This flow of universal time is conceived to be independent of space. We shall refer to such a space as the 'classical space' represented with the axes x,y,z perpendicular to each other:

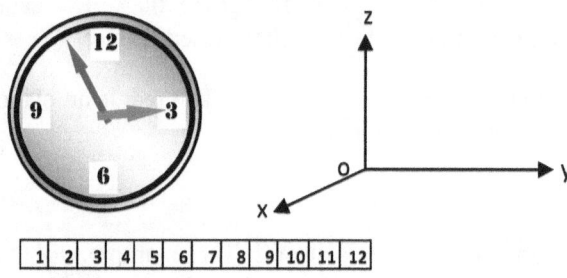

A Complex Space: A Mathematical Model of the Physical Space-Time as Conceived in Special Relativity

Two complex frames are used together. One: the primed frame, (x', ict') for the proper frame of a rigid rod and a clock. Two: the unprimed frame, (x, ict) as a proper frame for an observer. Here c is the speed of light, 300,000 km per second and $i \equiv \sqrt{-1}$. These complex frames are placed at a strange angle ψ clockwise to represent a relative velocity v of the rigid rod and clock with respect to the observer (tan ψ = v/ic).

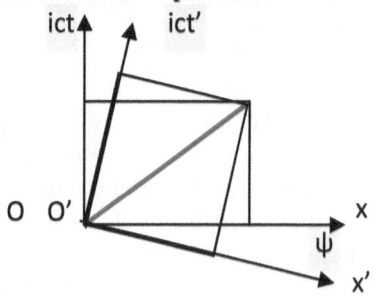

Complex space-time (x, it) and (x', it')

A Conceptual Error in the Calculation of Length Contraction In Special Relativity

First Limitation Imposed by Time on Our Perception of Space

Light travels at a fixed speed of 300,000 km per second which is same as 1 foot in a nanosecond, or one meter in 3 nanoseconds, or 300 meters in a microsecond or a kilometer in 3.33 microseconds.

A moment's reflection tells us that our perception of a length in reality is necessarily accompanied with a component of time. The star Betelgeuse we see in the Orion constellation in the night sky is already 640 years older. The sun we see in day time is already 8 minutes older. The moon we see is already 1.5 seconds older. When I notice you eight feet away, you are already 8 nanoseconds older. When I see along a bridge one kilometer long, the perception of the length is always accompanied with the passage of 3.33 microseconds. When I see the fingertips at the end of my stretched hand, the finger tips are already three nanoseconds older. When I hold a foot scale in my hand and see it edgewise I not only see a span of space twelve inches long, I also 'see' a span of time a nanosecond long. That I do not realize this fact is an illusion. But the image of the far end of the foot scale reaching my retina is a nanosecond older than the image of the near end of the foot scale reaching my retina. This is a physical fact. My perception of a span of space is always, necessarily, accompanied with the perception of a historical span of time. This is the *first* of the two influences of time on our perception of space. Perception of space is always inseparable from perception of an accompanying component of historical time whether we ordinarily realize this fact of physics or not.

The Physical Space-Time:

Every perception of the physical space by an observer is historical; this is the first of the two limitations imposed by time on our perception of space. Note the (-) signs in the figure below:

A Conceptual Error in the Calculation of Length Contraction In Special Relativity

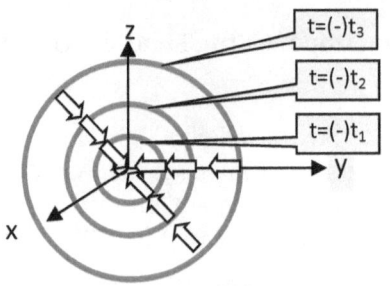

Classical space (x,y,z)

The light spheres carrying historical signals are collapsing on the observer at a speed of 300,000 km per second.

We can represent the above details in a simpler looking mathematical model but which is a little more sophisticated. We generally pay the price of sophistication for making mathematics a little simpler. We express the velocity of light, 300,000 km per second, as a unit of speed, $c = 1$. Now t shall represent a length ct, and v will represent the ratio of the velocities v/c, a unit less number (a ratio of two velocities).

The (x, it) Mathematical Space-Time:

We construct a mathematical space to represent the physical space with a fixed speed of light. The mathematical space is constructed with two Cartesian coordinates: (i) an axis-x for space and (ii) an axis-it for time. The space and time axes are orthogonal to each other. The space axes y and z are treated as suppressed. An observer is placed at the origin O of the (x, it) coordinates:

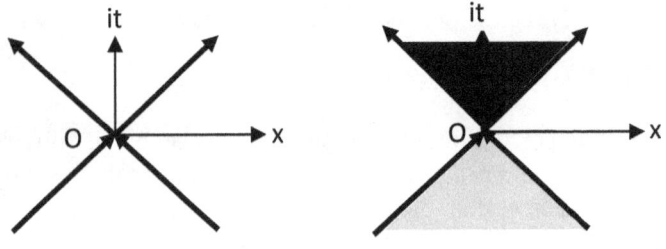

Complex space-time (x, it)

A Conceptual Error in the Calculation of Length Contraction In Special Relativity

We do not perceive the future. Our world experience is entirely confined to the past. It is, hence, a little more meaningful to draw the complex plane as a single light triangle in the past:

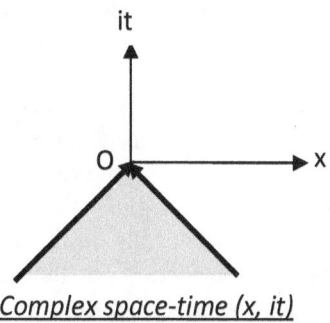

Complex space-time (x, it)

The shaded portion in the bottom half of the complex space-time (x, it) corresponds to the whole of the physical world of experience in the classical space (x, y, z). This, at any moment of time, is composed entirely of the experiences of the past.

This mathematical space (x, it) has some interesting peculiarities. The entire world of experience of the observer in the physical space is contained within the vertical spaces between the two 45^0 lines for which $|\Delta t| \geq |\Delta x|$. This mathematical space, corresponding to the world experience, is the vertical colored portions which we shall refer to as the light triangles and the borders of the light triangles as the light lines. The light lines are the trajectories of light in the space-time represented in the mathematical space above.

A very important realization:

It is of vital importance to realize that the phenomenon of light travelling along the physical x-axis from the positive x-direction to the observer placed at the origin (see the figure *Classical Space (x,y,z)* in the page -62-) is equivalent to the following three components of the complex (x, it) space (see the figure *Complex space-time (x, it)* above) taken together:

(i) the mathematical x-axis in the complex space,
(ii) the 45^0 line in the right side from the past   and
(iii) light travelling along this 45^0 line.

Peculiarity-One of the (x, it) Mathematical Space:

No single point of the physical space is represented outside the light triangles of the mathematical space above. Indeed the familiar x-axis of the classical space, a portion of the physical space, perceived by an observer at the origin cannot lie outside the light triangles, certainly not along the x-axis of the (x, it) mathematical space. Yes, what falls on the x-axis of the (x, it) space is a mathematical projection of the perceived x-axis of the classical (x,y,z) space. But this projection is just a mathematical, geometrical projection; it does not represent a physical entity, a physical length along the x-axis in the physical space . A mathematical projection can be outside the light triangle. But a physical entity is always within the light triangle. We must realize that the possibility of the observer at the point of origin somehow reaching out in to this outside-light-triangle-region by using a rigid rod is a big no-no, the possibility does not exist. Super luminal speeds are required for moving a rigid rod from inside to the outside of the light triangle. Infinite forces and energies shall be insufficient to achieve this purpose.

Peculiarity-Two of the (x, it) Mathematical Space

Another peculiarity of this mathematical space is an asymmetry between x-axis and it-axis : that the it-axis lies entirely within the light triangles whereas the x-axis is entirely outside except for the point of origin. An implication is this. The observer at the origin of (x, it) coordinates can perceive the passage of a *'pure'* duration of time, unaccompanied with a space projection. However, he cannot experience a *'pure'* span of length unaccompanied with duration of time. This is because, to be part of the observer's world experience, the span of length must lie within the light triangles which is possible only in accompaniment with a duration of time such that $|\Delta t| \geq |\Delta x|$; this condition must hold for a legitimate space-time interval.

Peculiarity-Three of the (x, it) Mathematical Space

Time changes in an unidirectional manner but space can be traversed in either direction for each of x, y, z axes. The space and the time projections, however, adjust and exchange with each other in order to keep the space-time interval an invariant amongst inertial observers.

Indeed, in writing $(-)\Delta x$, the negative sign is merely an algebraic convention. This does not affect the physical significance of space. One may travel from east to west as easily from west to east. However, in case of time, $(-)i\Delta t$ is not merely an algebraic convention but necessarily represents the physical fact that the duration of time experienced always belongs to the past and never to the future. One can go only in to the future and never in to the past.

Peculiarity-Four of the (x, it) Mathematical Space

The apparent length $\Delta x^2 + \Delta t^2$ of an interval is quite different from its actual length calculated in the complex plane: $\Delta x^2 + \Delta (it)^2 = \Delta x^2 - \Delta t^2$. Indeed, for a space-time interval lying on a light line, i.e., along a 45^0 line, its length in the complex plane is zero irrespective of its apparent length because $\Delta x^2 + \Delta (it)^2 \equiv \Delta x^2 - \Delta t^2 = 0$. One must always check by calculation the length of a line in the complex space and not get carried away by the apparent length.

In line with the last section we may write $\Delta x^2 + \Delta (-it)^2 = \Delta x^2 - \Delta t^2$.

The Central Postulate of Special Relativity

Starting with the postulate of the constancy of the speed of light in special relativity and making use of the generalization of $dx^2+dy^2+dz^2-c^2dt^2 = 0$ to $dx^2+dy^2+dz^2-c^2dt^2 = ds^2$ leads one to the transformation of coordinates of Hendrik Lorentz.

A Conceptual Error in the Calculation of Length Contraction In Special Relativity

Using $c = 1$ unit of velocity, the generalized equation is: $dx^2+dy^2+dz^2-dt^2 = ds^2$ where ds is defined as the space-time interval. Note, here t is a distance ct with $c = 1$.

The central postulate of special relativity is this: the generalized space-time interval between two events (x_1,y_1,z_1,t_1) and (x_2,y_2,z_2,t_2) is the same for all inertial observers: s_2-s_1 or Δs or ds is the same for all inertial observers. The standard language expressing this idea is: the space-time interval, ds, is an invariant for all inertial observers. The values of dx, dy, dz and dt adjust amongst themselves suitably in order to keep the value of the space-time interval ds the same for all the inertial observers.

The equations of transformation of coordinates between two inertial observers moving at a very high relative velocity with each other were first given by Hendrik Lorentz.

Hendrik Lorentz Transformation of Coordinates

For making a proper choice of mathematical framework, the two complex frames we express the space-time interval as $dx^2+dy^2+dz^2+(icdt)^2 = ds^2$. Here $i \equiv \sqrt{(-1)}$.

Now we make (x, ict), i.e., (x, it) our choice of coordinates and place an observer at the origin. For convenience we define the letter $l \equiv ct = t$ and, for further convenience, express all velocities of objects, coordinate frames, etc. in terms of units of c. Thus we express the space-time interval as $dx^2+dy^2+dz^2+(icdt)^2 = dx^2+dy^2+dz^2+(idt)^2 = ds^2$. We define below a complex frame as in the figure:

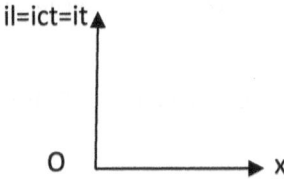

Complex space-time (x, it)

A Conceptual Error in the Calculation of Length Contraction In Special Relativity

For consistency in our discussions we shall limit ourselves to the pair of primed (x', it') and unprimed (x, it) complex frames.

The primed complex frame is chosen to be the proper frame of a rod and a clock moving at a high relative speed with respect to an observer in the unprimed frame.

The primed complex frame is placed at an angle ψ clockwise with respect to the unprimed complex frame to help us derive the equations of transformation of coordinates. The angle ψ represents the relative velocity in classical space. The idea shall be clear from the relationship $\tan \psi = v/ic$ appearing in the equation (2) coming later on. The origins of the two complex frames are made to coincide.

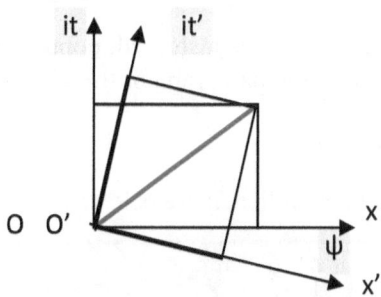

Complex space-time (x, it) and (x', it')

We make two further observations:

1. The rotation of the primed complex plane with respect to the unprimed complex frame is rather a very peculiar idea, even going by the standards of the familiar complex analysis. [Remember the fact that when a complex function $w = u(x,y) + iv(x,y)$ is taken up for the purpose of determining its analyticity, the two complex planes (x, iy) and (u, iv) are always kept apart and never merged together or rotated with respect to each other].

A Conceptual Error in the Calculation of Length Contraction In Special Relativity

2. In view of the foregoing discussions we must speak of a 'relativistic change of length' always along with the accompanying 'relativistic change of time'. Both the space and time projections must always be required to satisfy the following two cardinal principles:

(i) $\Delta x'^2 + \Delta(it')^2 = \Delta x^2 + \Delta(it)^2$, invariance of interval true for every interval in the complex planes, and,

(ii) $|\Delta t'| \geq |\Delta x'|$; $|\Delta t| \geq |\Delta x|$, true only for legitimate and physically sensible intervals within the light triangles, and not true for the superluminal regions outside the light triangles.

Errors in Calculations of Observed Length in Some Books

A list of errors in various historical, popular and text books is given in the concluding paragraph of this chapter.

Correct Equations

We reproduce the earlier figure below:

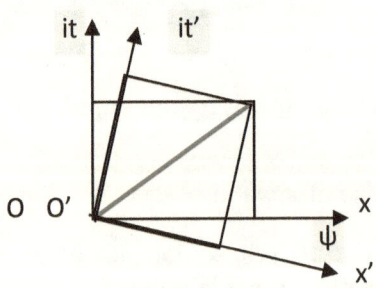

Complex space-time (x, it) and (x', it')

We write the following equations for the two mathematical projections of the rigid rod in its proper frame, the primed frame on x'-axis and it'-axis respectively:

A Conceptual Error in the Calculation of Length Contraction In Special Relativity

$$\left.\begin{array}{l}\Delta x' = \Delta x \cos \psi - i\Delta t \sin \psi \\ i\Delta t' = \Delta x \sin \psi + i\Delta t \cos \psi\end{array}\right\} \quad (1)$$

We consider a point fixed in the classical (x',y',z') space. For this point, $\Delta x' = 0$ in the (x', it') mathematical space. Thus, from equations (1), we receive:

$$\left.\begin{array}{l}\Delta x \cos \psi - i\Delta t \sin \psi = 0 \quad \text{i.e.,} \\ \tan \psi = v/ic \\ \cos \psi = \gamma > 1 \\ \sin \psi = (v/ic)\gamma \quad \text{where} \\ \gamma = 1/\sqrt{[1 - v^2/c^2]}\end{array}\right\} \quad (2)$$

Inserting various values from (2) into (1) and, writing velocity v in units of c, and simplifying, we receive the values of the two mathematical projections of the rigid rod as:

$$\left.\begin{array}{l}\Delta x' = [\Delta x - v\Delta t]/\sqrt{[1 - v^2]} \quad \text{and} \\ \Delta t' = [\Delta t - v\Delta x]/\sqrt{[1 - v^2]}\end{array}\right\} \quad (3)$$

We may solve the two equations (3) for Δx and Δt, mathematical projections of the rigid rod in the unprimed frame (x, it) and receive

$$\left.\begin{array}{l}\Delta x = [\Delta x' + v\Delta t']/\sqrt{[1 - v^2]} \quad \text{and} \\ \Delta t = [v\Delta x' + \Delta t']/\sqrt{[1 - v^2]}\end{array}\right\} \quad (4)$$

These are the familiar (Hendrik) Lorentz transformations.

The change of the observed length of the rigid rod from its proper length is the *second* of the double influence of time on perception of space.

A Conceptual Error in the Calculation of Length Contraction In Special Relativity

Untenable Consequence of Proposing an Unphysical Rod Along the x'-axis (with Zero Component Along the t'-axis):

We may, now, focus our attention on the consequence of the commonplace erroneous proposal of placing a rigid rod on the x'-axis of the primed frame (x', it') and assuming that $\Delta t' = 0$, i.e., erroneously treating the portion from (0, i0) to (x', i0) as a physical interval occupied by a rigid rod. To show the error, we insert $\Delta t' = 0$ in equations (4). We, thus, receive $\Delta x = [\Delta x']/\sqrt{[1 - v^2]}$ and $\Delta t = [v\Delta x']/\sqrt{[1 - v^2]}$. Considering Δt and Δx as positive quantities, and dividing Δt by Δx we further receive, $\Delta t/\Delta x = v < 1$ in units of c, i.e, $|\Delta t| < |\Delta x|$. This space-time interval in the unprimed frame is unphysical.

Summary

We reproduce the equations (4) below:

$$\left. \begin{array}{l} \Delta x = [\Delta x' + v\Delta t']/\sqrt{[1 - v^2]} \text{ and} \\ \Delta t = [v\Delta x' + \Delta t']/\sqrt{[1 - v^2]} \end{array} \right\} \quad (4)$$

We write L for the mathematical projection Δx, T for Δt and similarly L' and T' for the primed mathematical projections $\Delta x'$ and $\Delta t'$ and receive:

$$\left. \begin{array}{l} L = \gamma(L' + vT') \quad \text{and} \\ T = \gamma(vL' + T') \end{array} \right\} \quad (5)$$

Inverting, we receive

$$\left. \begin{array}{l} L' = \gamma(L - vT) \quad \text{and} \\ T' = \gamma(T - vL) \end{array} \right\} \quad (6)$$

In matrix form the system of equations (5) is:

$$\begin{bmatrix} L \\ T \end{bmatrix} = \begin{bmatrix} \gamma & \gamma v \\ \gamma v & \gamma \end{bmatrix} \begin{bmatrix} L' \\ T' \end{bmatrix} \quad (7)$$

A Conceptual Error in the Calculation of Length Contraction In Special Relativity

We verify that $L'^2 + (iT')^2 = L^2 + (iT)^2$. This satisfies the first cardinal principle.

On inverting the matrix equation we receive

$$\begin{bmatrix} L' \\ T' \end{bmatrix} = \begin{bmatrix} \gamma & (-)\gamma v \\ (-)\gamma v & \gamma \end{bmatrix} \begin{bmatrix} L \\ T \end{bmatrix} \qquad (8)$$

This corresponds to the system of equations (6).

These are the correct equations. Particularly,

$L = \gamma(L' + vT')$ (correct).

$L = \gamma L'$ (wrong).

The conceptual error here consists in treating $\gamma(vT')$ erroneously as being identically equal to zero.

Consider, for example, in a given equation: $A = B + C$, the term C [here $\gamma(vT')$] is not identically zero. If we remove this term C from the equation, an inequality: $A \neq B$ will result, not another equation.

We shall, next, explore the consequences of the correct equations on length alteration. Two tables are provided in the next two pages to bring out the interesting consequences.

A Conceptual Error in the Calculation of Length Contraction In Special Relativity

TABLE for *wrong* approach

Wrong approach

If T were zero, i.e., Δt were zero, i.e., t is a fixed moment of time, definite time, then:

$L' = \gamma L$ (wrong)

$L'/L = \gamma > 1$, contraction from proper length L' to L, observed length.

Wrong. T is never zero for a rigid rod.

L'	T	$\|T\| > \|L\|$	$S'^2 = L^2 - T^2$		v	v^2	$1-v^2$	$\sqrt{(1-v^2)}$	γ	$L = \gamma L' + \gamma vT$	$T = \gamma vL' + \gamma T$	$\|T\| > \|L\|$	$S'^2 = L^2 - T^2$		$S'^2 = S^2$?	wrong $\gamma L' = \gamma L$	same as first colour mn?	wrong L'/L	Observed/ proper (wrong) L/L'	wrong L'>L or L<L'
0.10	0.20	Yes	-0.03		0.50	0.25	0.75	0.87	1.15	0.23	0.29	Yes	-0.03		Yes	0.27	No	1.15	0.87	Contraction (wrong)
0.20	0.30	Yes	-0.05		0.50	0.25	0.75	0.87	1.15	0.40	0.46	Yes	-0.05		Yes	0.47	No	1.15	0.87	Contraction (wrong)
0.10	-0.15	Yes	-0.01		0.50	0.25	0.75	0.87	1.15	0.03	-0.12	Yes	-0.01		Yes	0.03	No	1.15	0.87	Contraction (wrong)
0.20	-0.25	Yes	-0.02		0.50	0.25	0.75	0.87	1.15	0.09	-0.17	Yes	-0.02		Yes	0.10	No	1.15	0.87	Contraction (wrong)

(negative T for the *past*)

A Conceptual Error in the Calculation of Length Contraction In Special Relativity

TABLE for correct approach

Correct approach
For a rigid rod, T is always > 0. Particularly, T ≥ L.
Thus,
L' = γ(L − vT) shall be used with a non-zero T. *(Correct)*.
L'/L = γ[1 − v(T/L)]

proper L'	proper T	|T'| ≥ |L'|	$S'^2 = L'^2 - T'^2$	v	v^2	$1-v^2$	$\sqrt{(1-v^2)}$	γ	observed L = γL' + γvT	observed T = γvL' + γT	|T| ≥ |L|	$S^2 = L^2 - T^2$	$S'^2 = S^2$?	correct L' = γL' − γvT	same as first mm?	correct L'/L	observed/proper (correct) L/L'; (dilation>1)	correct L'>L or L<L'
0.10	0.20	Yes	−0.03	0.50	0.25	0.75	0.87	1.15	0.23	0.29	Yes	−0.03	Yes	0.10	Yes	0.43	2.31	DILATION (correct)
0.20	0.30	Yes	−0.05	0.50	0.25	0.75	0.87	1.15	0.40	0.46	Yes	−0.05	Yes	0.20	Yes	0.49	2.02	DILATION (correct)
0.10	−0.15	Yes	−0.01	0.50	0.25	0.75	0.87	1.15	0.03	−0.12	Yes	−0.01	Yes	0.10	Yes	3.46	0.29	Contraction (correct)
0.20	−0.25	Yes	−0.02	0.50	0.25	0.75	0.87	1.15	0.09	−0.17	Yes	−0.02	Yes	0.20	Yes	2.31	0.43	Contraction (correct)

(negative T for the *past*)

A Conceptual Error in the Calculation of Length Contraction In Special Relativity

Conclusion: Errors in calculations of observed length in various historical, popular and text books

One book:

At definite time $t = 0$ *in the unprimed frame, the position of the integer* $x' = 1$ *is, with respect to the unprimed frame given by* $x = \sqrt{(1 - v^2)}$.

Error: $x' = 1$ *is a mathematical projection on the primed frame* x'-*axis and is NOT a physical entity, the rigid rod. A mathematical projection can be outside the light triangle. But a physical entity is always within the light triangle, inalienable from its associated mathematical projection in time. For the rigid rod the projection on the time axis must be* $|t'| \geq 1$. *Thus, for the observer in the unprimed frame,* $x = \sqrt{(1 - v^2)}$ *represents contraction of only a mathematical projection* $x' = 1$ *and NOT that of a physical entity.*

Another book:

Some books write: $x'_2 = \gamma(x_2 - vt)$ *and* $x'_1 = \gamma(x_1 - vt)$ *for the two ends of the moving rod, and* $t = t_1 = t_2$.

The error is obvious.

Yet another book:

Length Contraction: $L = x_2 - x_1$ *is on the* x-*axis of unprimed frame. Since* $t'_2 = t'_1$, *we have* $dx' = \gamma\, dx$. *As* $\gamma > 1$, $dx < dx'$. *So we have length contraction.*

Time Dilation: *Two events take place at the same spot* $x'_2 = x'_1$ *in the primed frame;* $dt = \gamma dt'$.

Error: The author has chosen the unprimed frame for length contraction but has chosen the primed frame for time dilation.

A Conceptual Error in the Calculation of Length Contraction In Special Relativity

Velocity of Light and Some Related Figures For Reference

One nanosecond is to one second is the same as one second is to 31.7 years.

	Velocity of light, c
300,000	km/second
300,000,000	m/second
984,251,969	feet/second
0.98	feet/nanosecond,
this is nearly 1	feet/nanosecond
300	m/microsecond
3.33	microseconds/km

	Conversion Table
3.280840	feet/meter,
1.00E+06	microseconds/second
1.00E+09	nanoseconds/ second
1.00E+12	picoseconds/ second
1.00E+15	femtoseconds/second

Defining a New Property *'Scale'* of Non-Singular Square Matrices

How I Came Across This Idea

This idea occurred to me while playing with numerical examples of Hendrik Lorentz transformation. A new property of matrices, 'scale', as defined here, may be used to obtain a 'scaled down' matrix, with a unit determinant, from a given non-singular square matrix:

Scale

GIVEN A NON-SINGULAR SQUARE MATRIX we may define its *'scale'* as

'Scale'=(absolute value of the determinant)$^{1/rank}$

A given matrix may be divided by its *'scale'* as defined above in order to obtain its *scaled down matrix*.

For the *scaled down* matrix both (1) the *absolute value of determinant* and (2) the *scale* will be unities, i.e., 1 each; numerical examples are given below followed by a general treatment:

Example 1

Given Matrix		Det	Scale	Scaled Down Matrix		Det	Scale
10	20	-500	22.36	0.45	0.89	-1	1
40	30			1.79	1.34		

Inverse Matrix		Det	Scale	Scaled Down Matrix		Det	Scale
-0.06	0.04	-0.002	0.04	-1.34	0.89	-1	1
0.08	-0.02			1.79	-0.45		

Example 2

Given Matrix			Det	Scale	Scaled Down Matrix			Det	Scale
-3	6	-11	10	2.15	-1.39	2.78	-5.11	1	1
3	-4	6			1.39	-1.86	2.78		
4	-8	13			1.86	-3.71	6.03		

Inverse Matrix			Det	Scale	Scaled Down Matrix			Det	Scale
-0.4	1	-0.8	0.1	0.46	-0.86	2.15	-1.72	1	1
-1.5	0.5	-1.5			-3.23	1.08	-3.23		
-0.8	0	-0.6			-1.72	0	-1.29		

Example 3

Given Matrix		Det	Scale	Scaled Down Matrix		Det	Scale
1 + i2	0	-5+i10	√(-5+i10)	(1+i2)/√(-5+i10)	0	-1	1
0	3+i4			0	(3+i4)/√(-5+i10)		

Inverse Matrix		Det	Scale	Scaled Down Matrix		Det	Scale
(3+i4)/(-5+i10)	0	1/(-5+i10)	1/√(-5+i10)	(3+i4)/√(-5+i10)	0	-1	1
0	(1+i2)/(-5+i10)			0	(1+i2)/√(-5+i10)		

General Proof

A general proof for the assertion that a '*scaled down*' square non-singular matrix will have a determinant of value 1:

Defining a New Property *'Scale'* of Non-Singular Square Matrices

$$\text{Matrix A} = [a_{ij}] = \begin{bmatrix} a_{11} & \text{---} & \text{---} & \text{---} & \text{---} & a_{1r} \\ \text{---} & \text{---} & \text{---} & \text{---} & \text{---} & \text{---} \\ a_{r1} & \text{---} & \text{---} & \text{---} & \text{---} & a_{rr} \end{bmatrix} \text{ of rank } r$$

Determinant $D = \Sigma\, a_{ij}\, C^{ij}$, summation is over either i *or* j; elements are either real or complex or both.

Indicating a *'scaled down'* quantity by underscoring, i.e., *scaled down* matrix by $\underline{A}$, *scaled down* element by $\underline{a}_{ij}$, *scaled down* cofactor by $\underline{C}^{ij}$, *scaled down* determinant by $\underline{D}$, etc., we receive:

$$\underline{A} = (1/D^{1/r})[a_{ij}/D^{1/r}] = \begin{bmatrix} a_{11}/D^{1/r} & \text{---} & \text{---} & a_{1r}/D^{1/r} \\ \text{---} & \text{---} & \text{---} & \text{---} \\ a_{r1}/D^{1/r} & \text{---} & \text{---} & a_{rr}/D^{1/r} \end{bmatrix} \text{ of rank } r$$

Determinant of the *scaled down* matrix $\underline{D} = \Sigma\, \underline{a}_{ij}\, \underline{C}^{ij}$, ($\underline{C}^{ij}$ are the *scaled* cofactors); summation is over either i or j.

Each term of the *scaled down* determinant is a product of exactly r terms each of which will contribute a factor of $1/D^{1/r}$; this factor may be abstracted out to leave behind the value D of the original determinant. These r abstracted factors each of $1/D^{1/r}$ will, thus, together, contribute a factor of $1/D$ which will *scale down* the original determinant of value D to $D*(1/D) = 1$.

Please Note

A 'scaled down' matrix will NOT generally be an orthogonal or unitary matrix as may easily be verified by simple counter examples.

A New Approach to the Conditions for Analyticity of Functions of Complex Variables

How I Got This Idea

I was somewhat dissatisfied by the text book method of mechanical, sequential movements superimposed on the complex plane for setting up of the equations leading to the Cauchy-Riemann conditions. Once in every five years or so, I would revisit the topic. At one time I was, for some purpose, preoccupied with the concept of inexactness. Cauchy integral theorem, which is intimately connected with the Cauchy-Riemann conditions, seemed to be a statement on exactness. One thing led to another and I discovered the new approach from the foundations of differential calculus (namely, $M_y = N_x$ rule for exactness of $Mdx + Ndy$), and presented a paper in a national conference on mathematics sponsored by CSIR and Odisha Mathematical Society (OMS) in the year 2003.

PART 1

Derivation of (i) the Cauchy-Riemann Conditions for Analyticity and (ii) Cauchy's Integral Theorem from Examination of Exactness of wdz

EXAMINATION OF "wdz" for exactness, a new approach, helps one to arrive at the Cauchy-Riemann conditions for analyticity as well as to neatly tie up with Cauchy's integral theorem of $\oint wdz = 0$ for analytic w.

Exact Differentials

We know from real analysis that if, for a given expression $M(x,y)dx + N(x,y)dy$, $M_y = N_x$, then the expression is an exact differential.

A New Approach to the Conditions for Analyticity of Functions of Complex Variables

We shall apply the same rule to the complex function $w \equiv u(x,y) + iv(x,y)$. We shall see that the question "whether the function w is analytic" is closely related to the question "whether wdz is exact". This relationship will elegantly tie up with the Cauchy's integral theorem of $\oint wdz = 0$ for analytic w.

We begin with,

$wdz \equiv (u+iv)(dx+idy) \equiv (u+iv)dx + (iu-v)dy$.

If this is to be exact, then, by the $<M_y = N_x>$ rule, then $(u+iv)_y = (-v+iu)_x$.

Separating the real and imaginary parts we receive

(i) $\quad u_x = v_y$ and

(ii) $\quad u_y = (-)v_x$.

These are the familiar Cauchy-Riemann conditions for analyticity of the complex function w.

Thus, w is analytical if and only if wdz is exact. Now wdz is exact implies $\oint wdz = 0$. This is the famous Cauchy's integral theorem for analytical functions.

Conclusion

We have, thus, examined the complex differential wdz for exactness by applying the familiar $<M_y = N_x>$ rule, a new approach, and arrived at the famous Cauchy-Riemann conditions for analyticity. We have also immediately tied up neatly with the equally famous Cauchy's integral theorem of $\oint wdz = 0$ for analytic w.

A New Approach to the Conditions for Analyticity of Functions of Complex Variables

PART 2

Unsuitability of the Analytical Function $w \equiv u+iv$ for Representing Conservative Force Fields

Notwithstanding the fact that $\oint wdz = 0$ for analytic w, we shall find that analytical function $w \equiv u+iv$ is not suitable for representing conservative force fields. This is counter-intuitive.

Examination of Analytical $w \equiv u+iv$ for the Purpose of Representing Conservative Force Fields

We notice that $\oint wdz = 0$ (for an analytical function $w \equiv u+iv$ satisfying the Cauchy–Riemann conditions for analyticity) looks like a statement on a conservative force fields w. One would be tempted to equate $wdz \equiv wdx + iwdy$ with the differential of a hypothetical potential field ψ: $d\psi \equiv (\partial\psi/\partial x)dx + (\partial\psi/\partial y)dy$. If indeed the potential function ψ exists, then, equating the real and the imaginary parts separately, we receive, $(\partial\psi/\partial x) = w$ and $(\partial\psi/\partial y) = iw$. However we also notice that wdz is a vector and is, as such, not suitable for representing energy which is a scalar. Besides, in order to represent energy, the magnitude of wdz should be a product of (i) the component of the force along the displacement, $|w| \cos\theta$ (θ being the angle between the force and the displacement) and (ii) the displacement, $|dz|$, i.e., energy is $|w||dz| \cos\theta$. But this is generally not true as may easily be verified for example, for $w = u$ and $dz = idy$: considering that u and iv are orthogonal, the work done is zero. But in this example, $wdz = uidy$ (an imaginary quantity) $\neq 0$. Thus, we find that w is not suitable for representing a conservative force field.

Further Observations by way of a Conclusion

The artificial attempt at defining a dot product as $z_1 \circ z_2 \equiv x_1x_2 + y_1y_2$ results in the expression $w \circ dz \equiv (u+iv) \circ (dx+idy) \equiv udx + vdy$. This, if it were exact (in order to represent a conservative force field), would yield $u_y = v_x$ which would immediately contradict the 2nd Cauchy-Riemann condition for analyticity of w (namely, $u_y = -v_x$). Thus w is not suitable for representing a conservative force field. We must, thus, be content to use the concept of the exactness of wdz for an elegant derivation of the Cauchy–Riemann conditions for analyticity and for neatly tying up with the Cauchy's integral theorem: $\oint wdz = 0$ for analytical w.

For a non-analytical w, clearly, wdz is inexact. However, dw is always exact. This is because the differential of any single algebraic expression like w is always exact. As an example, consider the complex conjugate $z^* = x-iy$ which is nonanalytic, $z^*dz = (x-iy)dx + (ix+y)dy$ is inexact, but $dz^* = dx - idy$ is exact as may easily be verified.

For given u(x,y) and v(x,y), in Real Analysis, we define p = iu+jv and, in Complex Analysis, we define w = u+iv. Note, if for a given choice of (u,v), wdz is exact, then, podr is inexact and if podr is exact, then wdz is inexact. These facts may easily be verified by applying the familiar $<M_y = N_x>$ rule for exactness.

Inadvisability of Import of Real Vector Concepts by Superficial Appearances into Complex Analysis

How I Came Around to This Idea

I sensed that something was wrong when I pored over the fact that the concepts of dot *and* **cross** *products,* **grad**, **curl**, div, laplacian, *etc. which have already been defined once in real analysis, are, once again, being defined in complex analysis. I wondered how one can define a thing twice in the same field of analysis. I suspected that there would be inconsistencies and found some. The definitions in complex analysis seem to have been adopted with the sole intention of keeping up superficial appearances with the definitions in real analysis. A concept cannot be defined in two different ways in two different places in the field of mathematics without risking consistency.*

Import of the Gradient

REAL ANALYSIS

GRADIENT, a vector, is defined on a scalar potential field, say, $S(x,y)$,

$$\mathbf{grad}\ S \equiv \mathbf{i}\ \partial S/\partial x + \mathbf{j}\ \partial S/\partial y.$$

COMPLEX ANALYSIS

Gradient, a vector, is presently defined on a complex quantity $w(x,y) \equiv u(x,y) + iv(x,y)$, a vector (unlike in the Real Analysis) as:

$\mathbf{grad}\ w \equiv (\partial/\partial x + i\partial/\partial y)(u + iv)$
$\equiv (u_x - v_y) + i(u_y + v_x)$ which is vector.

Some advantage might be seen in this definition of gradient as, equating the gradient to zero will immediately yield the Cauchy-Riemann conditions for the analyticity of w.

However one disadvantage is that, something else takes place in Real Analysis: when the gradient of $S(x,y)$ is set to zero, $S(x,y)$ turns out to be identically equal to a constant. It is much more elegant, hence, to consider the exactness of wdz for arriving at the Cauchy-Riemann conditions for analyticity of w as has been done in the essay "A New Approach to the Conditions for Analyticity of Complex Functions". Further, in Real Analysis, gradient of a vector would result in a dyadic and not a vector:

grad (**i**u + **j**v)

$\equiv$ **i**∂(**i**u + **j**v)/∂x + **j**∂(**i**u + **j**v)/∂y

$\equiv$ **ii**∂u/∂x + **ij**∂v/∂x + **ji**∂u/∂y + **jj**∂v/∂y . This is a dyadic and not a vector.

Import of the Vector Products

REAL ANALYSIS

Dot Product: **i** o **i** $\equiv$ 1,
< **j**, **k** similarly >;

Cross Product: $\quad$ **i** $\times$ **j** $\equiv$ **k**,

< **j**, **k** similarly >;
$\quad\quad\quad\quad\quad$ **i** $\times$ **j** $\equiv$ 0
< **j**, **k** similarly >

COMPLEX ANALYSIS
Simple Product:
$\quad$ Defined by ii $\equiv$ (-)1

Inadvisability of Import of Real Vector Concepts by Superficial Appearances into Complex Analysis

Dot Product:

Defined as

$$z1 \circ z2 \equiv |z1| \, |z2| \cos\theta$$
$$\equiv x1x2 + y1y2 \; Re\{z1^*z2\} \; ;$$

a real quantity.

This definition is similar in appearance to the definition of dot product in the Real Analysis. Under this definition we obtain:

$1 \circ 1 \equiv 1$,

$1 \circ i \equiv 0$,

$i \circ 1 \equiv 0$ and

$i \circ i \equiv 1$; note, in contrast, the simple product yields $ii \equiv (-)1$.

Exactness of $w \circ dz$ vs. That of $w \, dz$:

If we examine the exactness of $w \circ dz \equiv udx + vdy$ we obtain $u_y = v_x$ which immediately violates the 2nd of the Cauchy-Riemann conditions, namely, $u_y = (-)v_x$ implying that w is NOT analytical. Thus, if wdz is exact, then, $w \circ dz$ is NOT exact and vice- versa.

Cross Product:

Defined as

$$z1 \times z2 \equiv |z1| \, |z2| \sin\theta$$

$$\equiv x1y2 - y1x2 \equiv Im\{z1^*z2\};$$

is also a real quantity. The appearance, once again, is same as in Real Analysis. Under this definition we obtain

$1 \times 1 \equiv 0$

$1 \times i \equiv 1$, note difference with the real vector $\times$ product $< i \times j \equiv k >$;

$i \times 1 \equiv (-)1$, note difference with the real vector $\times$ product; and

$i \times i \equiv 0$; note, in contrast, the simple product yields $i\,i \equiv (-)1$.

Import of the Laplacian

REAL ANALYSIS

Laplacian, a scalar operator, is defined on a scalar, say, $S(x,y)$ as

$\nabla^2 S \equiv \text{div } \mathbf{grad}\ S \equiv \partial^2 S/\partial x^2 + \partial^2 S/\partial y^2$

COMPLEX ANALYSIS

Laplacian, a scalar operator, is defined on a vector $w(x,y)$ as

$\nabla^2 w \equiv \text{Re}\,\{(\text{del bar})(\text{del})\}\,w \equiv (\partial^2/\partial x^2 + \partial^2/\partial y^2)w$
$\equiv \partial^2 w/\partial x^2 + \partial^2 w/\partial y^2 \equiv \partial^2 u/\partial x^2 + \partial^2 u/\partial y^2 + i[\partial^2 v/\partial x^2 + \partial^2 v/\partial y^2]$.

Note, the definition of the laplacian in Complex analysis, notwithstanding the effort at maintaining appearances, is NOT the same as in Real Analysis. (If we force the Real Analysis definition of the laplacian, namely div **grad** on w, this will yield 0 for analytical w as **grad** $w = 0$ for analytical w. If w is not analytical, **grad** $w \equiv (u_x - v_y) + i(u_y + v_x)$ and we have, div **grad** $w \equiv (\partial/\partial_x\,(u_x - v_y) + \partial/\partial_y\,(u_y + v_x) \equiv u_{xx} - v_{yx} + u_{yy} + v_{xy} \equiv u_{xx} + u_{yy} \equiv \nabla^2 u \neq \nabla^2 w$

[Note, in Complex Analysis, divergence is defined as: div $(P + iQ) \equiv P_x + Q_y$ so as to keep up superficial appearances].

Thus the definitions in Real and Complex Analyses are not compatible with each other notwithstanding the appearances).

Remedy

The concepts of dot and cross products, **grad, curl**, div, laplacian, etc. as defined in complex analysis, do not go coherently together with the definitions of the concepts in the real analysis.

The remedy is to abandon the complex analysis definitions of dot product, cross product, **grad, curl,** div, laplacian, etc. and leave the subject alone with its virgin beauty.

Review of the Complex Potential

Why Was I Worried About 'Velocity Potential'?

I could find nothing 'actual' corresponding to the 'potential' in the term 'velocity potential'. Later on I found corroboration when I was going through Feynman's Lectures in Physics.

Complex Potential

IN THE FOLLOWING APPLICATIONS relating to the complex potential the considerations of analyticity are quite sufficient; it is not necessary to invoke the laplacian. We determine, for a given function u(x,y), its conjugate v(x,y) by using the Cauchy-Riemann conditions and, with these scalar functions, form the complex potential **w** ≡ u+iv which is a vector and is analytical (satisfying the Cauchy-Riemann conditions). This enables us to apply the technique of conformal mapping which immediately tells us that the equations "u(x,y) = constant" and "v(x,y) = constant" will yield lines orthogonal to each other.

Clearly "(real) **grad** u" lines and "v(x,y) = constant" lines are parallel to each other, each being orthogonal to the "u(x,y) = constant" lines. This is further corroborated by the following observation: since **w** is analytical the Cauchy-Riemann conditions apply which are $u_x = v_y$ and $u_y = (-)v_x$. Now (real) **grad** v ≡ **i**v_x+**j**v_y. Substituting the Cauchy-Riemann conditions to convert v - terms in to u - terms we obtain (real) **grad** v ≡ **i**(-)u_y+**j**u_x. As can be seen, the (real) dot product between (real) **grad** u and (real) **grad** v yields zero indicating that the (real) **grad** u and (real) **grad** v vectors are orthogonal to each other. This is quite expected.

It is customary to interpret u(x,y), a scalar, as a real potential function (yielding equipotential lines by the equation "u(x, y) = constant") and **w**, a vector, as the corresponding complex potential function. Now, there are two different interpretations available regarding v(x,y) which are discussed below:

Interpretation of v(x,y) as Yielding Lines of Force

One of the interpretations is to treat the equation "v(x, y) = constant" yielding lines of force. It may be noted that whereas (real) **grad** u , a vector, gives the magnitude of the force at the point (x, y) , the equation "v(x, y) = constant" gives the lines of force; (real) **grad** u is tangential to the lines "v(x,y) = constant" . This interpretation is comprehensive and is quite satisfactory.

Interpretation of v(x,y) Yielding Stream Lines

Real Analysis

Velocity is defined as $\mathbf{V}(x,y) = \mathbf{i}V_1(x,y) + \mathbf{j}V_2(x,y)$. The flow is supposed to be irrotational, (real) **curl V = 0** ; incompressible, (real) div **V** = 0 ; free from sources and sinks; and non-viscous. As the flow is irrotational, $V_1 dx + V_2 dy$ is exact, and is defined as du(x, y) , with $u_x \equiv V_1$ and $u_y \equiv V_2$. This u is sought to be treated as a "potential" coined as "velocity potential" though there is nothing physically "actual" to correspond to this "potential". To see how illogical and unfortunate is the nomenclature, consider the flow of a non-viscous, incompressible fluid in a large and long horizontal (x-axis) pipe line offering no resistance, the flow is without acceleration, smooth and laminar and there are no sources and sinks, i.e., where V_1 is constant and $V_2 = 0$; now $du \equiv V_1 dx$, i.e., $u(x,y) \equiv \int V_1 \, dx$; we find no "potential" physically corresponding to $\int V_1 \, dx$. Here is a case where a model, which is a mental object, is stretched beyond its reasonable limits of application, yielding a confusion of terms, here with "velocity" and "potential". One may not question the mathematician's freedom to define terms. But, here, it is important to keep in mind that there is nothing "actual", corresponding to the "potential" in the term "velocity potential", that ever manifests in any circumstance, (as it happens with potential energy that manifests as kinetic energy). It is not clear why a compulsion is felt to define a potential just because the du term is exact; there are many physical applications, say, for example, in thermodynamics where many exact differentials occur and it is not necessary to define a potential every time.

The following is an excerption from section 8, chapter 1 of GH Hardy's classic "A Course of Pure Mathematics" published by the English Language Book Society (ELBS) and Cambridge University Press: *"What is essential in mathematics is that its symbols should be capable of some interpretation; generally they are capable of many, and then, so far as mathematics is concerned, it does not matter which we adopt"*. Bertrand Russell has, further, said that *"**mathematics is the science in which we do not know what we are talking about, and do not care whether what we are talking about it is true**"*.

But a scientist or an engineer is always obliged to so interpret the mathematical terms that no confusion results and the treatment of a subject area is clear and coherent.

Complex Analysis

The velocity is, now, represented as $V = V_1+iV_2 = u_x+iu_y$. Since u, a scalar, is defined as a "potential", it is sought to be generalized by proposing a complex potential **w**, a vector. For this "potential" u(x,y) a complex conjugate v(x,y) is sought through use of the Cauchy-Riemann conditions and the complex potential is, now, defined as $w \equiv u+iv$. As the complex potential **w** is formed out of conjugates u & v, **w** is analytic (subject to existence of 2nd order PD, etc.). $w' \equiv dw/dz = dw/dx$ (by definition of analyticity) $\equiv u_x+iv_x = u_x-iu_y$, the latter relation is the 2nd Cauchy-Riemann condition for analyticity. Here is a woe: $w' = u_x - iu_y$ does not look like complex velocity V; this difficulty is sought to be circumvented by defining the complex velocity as $V \equiv V_1 + iV_2 \equiv u_x + iu_y \equiv w'^*$. Clearly V is not analytical as its conjugate w' is analytical. Here we have a complex potential **w**, which is analytical and a complex velocity V which is NOT analytical. It is not clear, what substantial advantages accrue (accompanied with many conceptual difficulties as presented above) over the Real Analysis treatment of potential functions in the treatment of generalized complex potential.

Remedy

The remedy, clearly, is to abandon the use of the term "velocity potential" as a misnomer. Such inconsistent terms are a hurdle to a student of mathematics and should be extirpated from mathematical literature. Another candidate term is "imaginary": what is more imaginary about $\sqrt{-1}$ apple compared to -1 apple?

Algebraic Solution of Chemical Equations

How I Got This Idea

I was pretty bad at chemistry. The subject matter was mysterious to me. I could not handle partial balancing, oxidation-reduction method, etc. for balancing chemical equations satisfactorily. It occurred to me that the number of atoms is the same for each element on either side of the chemical equation. Here I present an algebraic method for solving chemical equations using a set of linear algebraic equations generated from the given chemical equation. The method is easily adapted to matrices.

Introduction

OFTEN WE COME ACROSS chemical equations which are difficult to balance. This new approach is a logical extension of the so called trial and error method. The new method requires simultaneous solution of a set of linear algebraic equations.

The basic idea is that, the number of atoms of a given element remains same on either side of the chemical equation. The following example is used as an illustration.

EXAMPLE 1

$Sb + HNO_3 \Rightarrow H_3SbO_4 + NO_2 + H_2O$

We write

$\alpha Sb + \beta HNO_3 = \gamma H_3SbO_4 + \delta NO_2 + \varepsilon H_2O$

Balancing the number of atoms for each element on both the sides, we receive:

Algebraic Solution of Chemical Equations

Sb : $\quad \alpha = \gamma$

H : $\quad \beta = 3\gamma + 2\varepsilon$

N : $\quad \beta = \delta$

O : $\quad 3\beta = 4\gamma + 2\delta + \varepsilon$

Solving, we receive,

$\alpha = \alpha$

$\beta = 5\alpha$

$\gamma = \alpha$

$\delta = 5\alpha$

$\varepsilon = \alpha$

Assuming $\alpha = 1$, we receive,

$Sb + 5HNO_3 = H_3SbO_4 + 5NO_2 + H_2O$

A little experience helps solve the equations much quicker. For example, take

EXAMPLE 2

$Ca_2B_6O_{11} + SO_2 + H_2O \Rightarrow CaSO_3 + H_3BO_3$. We write,

$\alpha Ca_2B_6O_{11} + \beta SO_2 + \gamma H_2O = \beta CaSO_3 + 6\alpha H_3BO_3$.

(On the R.H.S. we wrote $\beta CaSO_3$, because 'S' appears on the L.H.S., and in the R.H.S. only in single terms. Hence, the co-efficient must be β. Similarly considering the number of atoms of 'B' on the L.H.S., we wrote $6\alpha H_3BO_3$ on the R.H.S.).

Algebraic Solution of Chemical Equations

Now let us write equations for Ca, O & H (since the number of atoms of S & B have already been taken into accounts when we wrote $CaSO_3$ and $6H_3BO_3$ in the R.H.S.):

$$Ca: \quad 2\alpha = \beta$$

$$O: \quad 11\alpha + 2\beta + \gamma = 3\beta + 18\alpha$$

$$H: \quad 2\gamma = 18\alpha$$

Assuming $\alpha = 1$ and solving,

$$\alpha = 1$$

$$\beta = 2\alpha = 2$$

$$\gamma = 9\alpha = 9$$

Thus, we receive,

$Ca_2B_6O_{11} + 2SO_2 + 9H_2O = 2CaSO_3 + 6H_3BO_3$

EXAMPLE 3

$\alpha KMnO_4 + \beta SO_2 + \gamma H_2O = (\alpha/2) K_2SO_4 + \alpha MnSO_4 + \delta H_2SO_4$

For S: $\quad \beta = \alpha/2 + \alpha + \delta$

O: $\quad 4\alpha + 2\beta + \gamma = 2\alpha + 4\alpha + 4\delta$

H: $\quad 2\gamma = \delta$

Assuming $\alpha = 2$ and solving,

$$\alpha = 2$$

$$\beta = (5/2)\alpha = 5$$

$$\gamma = \alpha = 2$$

$$\delta = 2.$$

Algebraic Solution of Chemical Equations

We, thus, receive:

$2KMnO_4 + 5SO_2 + 2H_2O = K_2SO_4 + 2MnSO_4 + 2H_2SO_4$

EXAMPLE 4

$2\alpha KOH + \beta I_2 = \gamma KI + \delta KIO_3 + \alpha H_2O$

(Note, H occurs only in one term on each side. To avoid $\alpha/2$ we write '2α' before KOH and α before H_2O).

For K : $\quad 2\alpha = \gamma + \delta$

O : $\quad 2\alpha = 3\delta + \alpha$

I : $\quad 2\beta = \gamma + \delta$

Assuming $\delta = 1$ and solving,

$\alpha = 3$

$\beta = 3$

$\gamma = 5$

$\delta = 1$

Thus, we receive

$6KOH + 3I_2 = 5KI + KIO_3 + 3H_2O$

EXAMPLE 5

$\alpha ICl + 3\beta H_2O = \gamma I_2 + \beta HIO_3 + \alpha HCl$

For I : $\quad \alpha = 2\gamma + \beta$
     H : $\quad 6\beta = \beta + \alpha$

Assuming $\beta = 1$, we receive

$$\alpha = 5\beta = 5$$

and $\gamma = 2\beta = 2$.

Thus, we receive,

$5ICl + 3H_2O = 2I_2 + HIO_3 + 5HCl$.

Conclusion

In general the number of unknowns in the resulting set of linear algebraic equations is one more than the number of equations. By assigning a simple small integer value to one of the unknowns it is quite easy to solve the set of equations. Today the same idea is available in the internet as chemical equation calculator in a user-friendly format.

Sustainable Development, Entropy, Madness and the Proposed Unit of Heat: _Hiroshima Atom Bomb Equivalent_ (HABE)

How I Got This Idea

Many years ago, I came across Radiative Forcing, RF, which is defined as the change in the net downward radiation (solar and infrared) at the tropopause, 10 Kilometers above the earth surface, because of greenhouse gases, resulting in global warming. RF is estimated at around 2.45 watts per square meter. The heat energy, thus, raining down upon the earth surface from the tropopause every hour is some 16,000 times the heat content of the atom bomb that was dropped on Hiroshima. We may define a unit of heat, <u>HABE</u>, as the heat content of the atom bomb which was dropped on Hiroshima. This unit of heat is to be used irrespective of radiation, explosion or any other effect of the atom bomb. We may, thus say, 16,000 HABEs of heat energy is raining down on the earth's surface every hour as a consequence of Radiative Forcing. Hiroshima is a powerful symbol of mass destruction in the modern times and may also be used as a meaningful unit of heat to represent the global warming. HABE, as a unit of heat will carry a visual impact that is absent in other units of heat such as the terajoue. Global warming is leading to mass extinction of species, which may, eventually, lead to the extinction of the human species as well. We may adopt as a remedial measure, capping and rolling back the bloating human civilization. This we may achieve by way of mass education, equitable distribution of wealth (education and prosperity being the best contraceptives) and total dependence on solar energy, etc.

Sustainable Development, Entropy, Madness and the Proposed Unit of Heat: Hiroshima Atom Bomb Equivalent (HABE)

IT IS SHOCKING TO REALIZE that the mankind is continuing to consume natural resources at an exponentially increasing rate, while, at the same time, piously making noises of saving the environment by "sustainable development" defined in a manner which resembles "eating a cake and having it too", a contradiction in terms. Looking at the developed nations whose debts exceed their respective national GDPs, ever widening deficit budgets and decoupling of currency from the gold standard, (and, further, with modern India and China, aiming to grow at 10 and 15% annually), all indicators of reckless consumption patterns of the earth's limited resources. This may quicken possible destruction of the earth as a habitable planet. World will do well to learn from the 5000 years of continuously living civilization and culture of India as a proven model of sustainability.

Story of the Game of Chess

The game of chess was invented in India. There is a beautiful anecdote regarding the invention of the game of chess. The inventor was a mathematician who presented his invention to his king. The king was very much fascinated and offered to reward him generously. But, to the chagrin of the king, the mathematician sought only some quantity of wheat as his reward. He wanted one grain of wheat in the first square, two grains in the second square, four in the third square, eight in the fourth square, doubling the quantity of wheat for every subsequent square on the chessboard. The angry king dismissed the mathematician from his sight and ordered the chief minister to hand over the quantity of wheat by next day morning. To his utter surprise, the chief minister informed the king next day morning that such a quantity of wheat does not exist on the planet earth. This is an example of how an exponential demand (read, conspicuous consumption) overwhelms the limited resources of the earth no matter how large.

Sustainable Development, Entropy, Madness and the Proposed Unit of Heat: Hiroshima Atom Bomb Equivalent (HABE)

<u>Law</u>: *No matter how large, no limited natural resource of a planet shall sustain an exponentially increasing demand on it for any great length of time.*

The rate of consumption of fossil fuel, for example, has increased exponentially from 4 to 10 GTOE in last fifty years. The population of the world has increased exponentially from 2 to 7 billion in the same period of time. Apart from the increase of population by nearly 4 fold, the individual's requirements have also increased, say, nearly 10 fold in the mean while. The atmospheric CO_2 has increased exponentially from 280 ppm to 350 ppm in the same period of time. The global warming has raised the average temperature of the world by about 0.5 degrees Celsius mostly within the last 50 years. The trend is exponential. Such a temperature rise had not taken place in the last 10,000 years of the geological history. *The heat energy involved in melting of ice to raise sea level by one millimeter is 1.5×10^{14} megajoules. This is 1.8 million HABEs. If we assume that the entire 100 mm of rise of sea level in the last 50 years is from melting of ice, then this represents nearly 50 million HABEs which is nearly an atom bomb for every twenty five people on the earth.*

'Hiroshima Atom Bomb Equivalent' (HABE) *Defined*

We pump in a million tons of carbon worldwide in to the atmosphere every hour i.e., a lac tons of carbon every 6 minutes or a lac tons of carbon dioxide every two minutes. This is giving rise to the enhanced greenhouse effect causing a rain of more than 4 times the heat content of the atom bomb - dropped over Hiroshima - from the stratosphere some 10 kilometers above to the earth's surface every second. Every year we commemorate the day Hiroshima was bombed. How shall we commemorate the 4 atom bombs' worth of heat raining down every second?

This has brought about the enhanced greenhouse effect which traps some of the heat that would otherwise radiate away in to space and sends it down to the earth's surface by way of

`RADIATIVE FORCING' or "RF". Let us define an unit of heat energy for the purpose of this essay in terms of the heat energy yielded by the atomic bomb exploded over Hiroshima which was 82 terajoules. Let us name this quantity of energy of 82 terajoules or as one unit of **HABE, Hiroshima Atomic Bomb Equivalent** of heat energy, to be used irrespective of the aspects of nuclear radiation or explosion. The heat energy that is raining down upon the earth surface from tropopause is about four HABEs every second!

The sea level worldwide has increased by around 100 mm because of global warming. Each mm rise in the sea level from melting of ice in Antarctica and Greenland represents enough heat to be equivalent to nearly 8 million atom bombs similar to the ones dropped over Hiroshima. Spread uniformly all over the globe this will be a bomb every 10 kilometers from each other. Here we consider only the heat as if accumulated gradually without any blast or nuclear effects.

Human population has grown exponentially under the favorable conditions brought about by engineering, technology, medicine which, at the same time, led to mass extinction of other species.

'Adequate Population' *Defined*

We would like to define, for the purpose of this essay, a population of 2 billion (year 1950) as an '**Adequate population**'. This population had given us our rockets and airplanes, radio and television sets, relatively and quantum mechanics, and had carried forward the rich arts, literature, philosophies and mathematics of the earlier times.

Entropy, Development and the Environment

The law of monotonous increase of entropy in isolated thermodynamic systems implies monotonous increase of disorder in the universe. When we bring in development to a location, we bring in order in the form of engineering designs, standards, quality, etc. This throws out a corresponding amount of disorder in to the neighboring environment in keeping with the law monotonous increase of entropy. When we use bulldozers and such other equipment operating on thermodynamic cycles, or electric motors that get their power ultimately from thermal power plants including the nuclear operating on thermodynamic cycles, they are all subject to the limitation of the Carnot's cycle which is the ideal thermodynamic cycle. Carnot's cycle practically ensures that we throw at least twice as much heat in to the surrounding environment as is converted in to useful work. This translates in to throwing an additional double amount of entropy, i.e., disorder in to the surrounding environment. Thus, when we bring in development to a location, we throw at least thrice times as much entropy, i.e., disorder, in to the surrounding environment as the order associated with the development.

[Entropy and Mind

I am sorry what I am going to touch upon here may appear out of context. However, I am unable to resist myself. It appears to me that we may be able to apply the concept of thermodynamic entropy conceptually to our mind as well. The idea is this: as we progressively discipline a part of our mind, we throw a correspondingly large amount of entropy and heat to the rest of it. Look at students furiously preparing for tests. Also think of the common citizen in more developed parts of the world where he has to train himself on ever growing amounts of skills, complexity and coordination of information. This is like walking several tight ropes all at the same time. No wonder, more people are stressed the more a civilization is developed. Perhaps, the concept of Gross Domestic Happiness adopted by Bhutan is superior to GDP.]

Large Number of Emerging Areas of Studies and Failed Technologies

We may interpret the large number of studies emerging in various areas in the last few decades as indicative of the desperation of the mankind to catch up with the large number of woes emerging from large scale interference with the environment. Exactly what is the score card as of now? If the continuing problems of the environment and the number of failed technologies are any indication, we are not doing well, nor going to do well in the foreseeable future. Consider the list of various new areas of studies first. Last twenty years saw the following partial list of studies emerge in the area of industrial maintenance: (i) breakdown maintenance, (ii) preventive maintenance, (iii) condition based maintenance, (iv) TPM, (v) six sigma quality, (vi) ISO 9k, etc., you too may add your endless lists from the legal, economic, administrative and management studies.

Next, the failed technologies! (i) Solar PV not increasing beyond 10-12% efficiency commercially for last 30 years, (ii) no superconductors carrying bulk power at zero loss, (iii) total cycle pollution load of fuel cell is no less on the environment, leading to pull back of research allocations worldwide, (iv) no new technology in the automotive sector in place of the piston and cylinder engines under use for the last 150 years (the Wankel engine did not take off commercially), (v) no fusion power that was promised decades ago, (vi) no economic solution is available for storage of electricity, (vii) no economic solution found for corrosion resistance in building metals.

Let us not be unduly optimistic of our technological puissance to fix any and all problems emerging out of very large scale anthropogenic disturbance of the natural environment. Clearly the speed of destruction far outstrips the speed of proven technological quick fixes.

A Recipe for Escaping the Global Warming Mess

Not easy. Organizations responsible to the world should actually quantify, certify or verify that the technical or other measures being adopted in the name of sustainable development are indeed adequate to meet global goals of sustainability, one of which is elimination of global temperature rise with in a decade. Citizens should articulate such demands to responsible organizations. (1) cap and roll back the civilization, scale down the civilization to a population of two billion, an 'adequate population' by adapting education and prosperity as the best contraceptives, (2) meanwhile, plan for executing a controlled crash of the human economy which is necessarily a small subset of the environmental economy, equitable distribution of wealth shall help the process (may be for a generation an educated population shall demand comforts at the cost of the environment, but soon afterwards, a reduced population shall reverse the effect), (3) switch over to solar and wind powers, a 100% carbon free power generation, (4) divert a vast proportion of the population to cultural (soft) activities rather than civilization building (hard) activities, (5) demand demonstrable verification and certification of sustainability for proposed technologies (for illustration, a proposed technology may be certified as 40% sustainable if it is demonstrated by acceptable methods of estimation or experimental demonstration that the technology will help reduce the contribution of greenhouse gases worldwide, emanating from the corresponding sector of application by 40% per year provided that the technology is adapted 70% world over in coming 10 years' time). By a slew of several such methods we may be able to defer an early destruction of our spaceship, the planet earth.

Story of a Madman (By Way Of Conclusion)

Once, while driving his car on a deserted road, a gentleman felt that the wheels are wobbling. Getting down from his car, he saw that all the four nuts of a front wheel are missing. While he was scratching his head, he sensed that a person was beckoning him from the other side of a barbed wire fence. The person advised him to remove a nut from each of the remaining three wheels and use these three nuts on the fourth wheel and drive on to the next town and complete the repairs. The gentleman thanked the person but, somehow, was puzzled. On, confirming that the barbed wire fencing encloses a lunatic asylum, he wondered aloud, how a mad fellow could offer a workable solution for repairing the car which did not occur to him. The inmate of the lunatic asylum told him that 'I may be mad but I am not an idiot.'

This is, more or less, the story of the human civilization, today, which has gone mad and is fast moving to a collective suicide, pushing the planet way beyond its carrying capacity. This form of madness is apart from the other well-known madness related to the accumulated capacity for thermonuclear destruction of the world many times over, MAD: mutually assured destruction! Our earth is a space ship, a sphere some twelve thousand kilometers across, hurtling through space some three million kilometers a day, complete with a fragile but adequate life supporting system which we are doing everything to destroy. The environment is so poisonous today that it no more supports the life of large animals like the jungle cats, vultures, and many other life forms in substantial numbers as it did earlier. Can time be very far away for the poison and destruction to boomerang on the Homo sapiens? Let us refrain from destroying our beautiful blue-green space ship in frenzy. There no other known planet like this within a hundred light years.

Exponential Behavior of Environmental Parameters

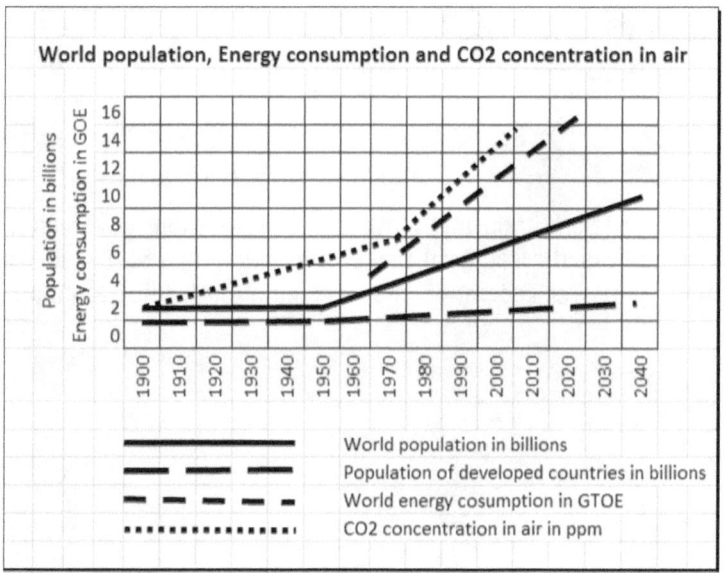

"At present, we are stealing the future, selling it in the present, and calling it GDP".

-Paul Hawken

Sustainable Development, Entropy, Madness and the Proposed Unit of Heat: Hiroshima Atom Bomb Equivalent (HABE)

The Root Cause of Environmental Disturbances

<u>Engineering and technology</u>
|
Population→Deforestation →Loss of bio diversity
| |→Soil erosion→ Earthquakes
|→Food, housing, clothing, transportation, governance, education, medicare, fossil fuel consumption on a vast scale for industry, power generation and transportation leading to energy crisis
|
Air pollution
|
Enhanced greenhouse effect
|
Global warming
|→Sea level rise-→ Earthquakes (partly) →Tsunami
Sun strokes, super cyclones, flash floods, etc.

> *"Gross National Happiness is more important than GDP".*
>
> - *HM Jigme Singye Wangchuk, King of Bhutan*

An Accounting Framework for Presenting Specific Energy and Energy Conservation Figures for an Industrial Plant

How I Got This Idea

I have seen many discussions on energy conservation where specific consumptions based on total production are discussed passionately and fruitlessly where explanations are demanded by those in authority and lame excuses offered by those who are not. Nobody realizes that there are large varieties of specific consumptions and these are not absolute quantities but are mere indices, that some are useful for setting up a project but wholly others are useful for saving money while operating the plant. Poring over all these aspects I developed an accounting framework for presenting specific energy and energy conservation figures for energy audit of an industrial plant.

IT IS NECESSARY TO FORMULATE an acceptable accounting framework for presenting specific energy and energy conservation figures for energy audit of an industrial plant. Energy conservation for an industrial plant must primarily answer the question: how much more we'd have had to pay extra for energy had we not undertaken certain energy conservation activities. There may be a dozen alternative choices for specific energies. A proper choice is very important as a wise choice shall render unnecessary many explanations in future.

Errors of Misrepresentation

The following two types of errors are likely in energy conservation estimates based on (i) year-end-results for plants having multiple products where (ii) specific energy is based solely on the dominant product (say product A):

Error of TYPE- I

As the specific energy is based on product A by choice, an increase in the production of a lesser product (say product B) may result in a larger consumption of energy in the year-end-result and may be erroneously perceived as a loss to the plant.

Error of TYPE- II

A decrease in the production of a lesser product (say product B) may result in a lesser consumption of energy in the year-end-result and may be erroneously perceived as a saving.

Thus, choice of specific energy with respect to one or more dominant products is NOT a good choice for estimating energy conservation figure for a facility/unit. Such a choice of specific energy cannot help us to trace the gain or loss to either efficient or otherwise operation of any equipment or systems of equipment. (However such a choice of specific energy based on one or a few dominant products may be useful for examining commercial viability of selecting one out of several alternative processes before establishment of a plant).

An accounting framework shall be acceptable for presenting specific energy and energy conservation figures for an industrial plant only if it is free or substantially free from both the types of errors I & II and, further, if the framework will provide figures that can be traceable to efficient or otherwise operation of equipment or system of equipment.

Issue 1: Additional Transport Equipment to Ensure Supply of an Input Material from a Larger Distance

Sometimes it may become inevitable to ensure supply of raw materials from a larger distance. Whereas this may speak on the commercial viability of a process, this has nothing to do with the efficient operation of installed equipment. This is a necessary expenditure of energy and does NOT constitute wastage of energy. Specific energy based solely on the dominant product (say product A) shall result in Error of TYPE-I.

Recommendation

Calculate specific energy after correcting for energy spent on extra transportation equipment.

Issue 2: Intermediate Buffer Stock

The energy expended to create an intermediate buffer stock shall inflate the year-end-result and should be deducted before specific energy is estimated based on the dominant product (say product A). Otherwise the result will be an Error of TYPE- I.

Recommendation

Calculate specific energy figure for each plant separated by the intermediate buffer stock. Also, this must be declared in the disclosure of practices in the accounting framework.

Issue 3: Correct Choice of Specific Energy

For plants having multiple products there will be a large variety of specific energies possible. This is because one may choose the basis from among many possibilities like (i) the dominant product, say A; (ii) simple sum of all the products, say $A + B + C + D + $ --; (iii) weighted sum of all or some of the products, $aA + bB + cC + dD + $ -- where a, b, c, d, etc. are the chosen weights; (iii) the weights may be chosen as the market price which may vary from year-to-year; (iv) the feed-to-plant; etc. Nearly every one of these will be asked for or used from time-to-time for various purposes of comparison or explanation. However specific energy figures based on one or more products in a multi-product plant will lead to loading these figures arbitrarily with overhead energy consumption from the products which are not represented in the basis for specific consumption). Even where all products are considered, the questions of (i) proper weightage and (ii) energy used for producing rejects, which is apt to be large (considering that waste shall generally travel from the beginning to the end of the process consuming energy all the way), are unsettling issues. Still more interesting is a particular reject which suddenly finds a value in the market or a product which is suddenly losing the market.

Recommendation
Choose specific energy on FEED- TO- PLANT basis.

AN ACCOUNTING FRAMEWORK

Financial Year Under Audit: 20__ - 20__

Energy Conservation

Summary Statement: Electricity/ FO/ Coal/Gas/Water, etc. (Separate Estimates)

Plant	Choice of analysis: (Prorata or nominal)
Plant 1	Prorata (based on chosen specific energy)
Plant 2	Prorata
Utility 1	Prorata
Utility 2	Prorata
Housing Colony	Nominal
(based on year-to-year difference)	
Admin. Building:	Nominal
	Total

Disclosures

1) Type of analysis:
(a) Prorata (based on specific consumption; suitable for large installations with measurable output),

(b) Nominal (yr-to-yr differences for small installations) as in the Summary statement above: Electricity/ FO/ Coal/Gas/Water, etc.

2) Plant wise practices:
a) Whether figures are presented after deducting energy expended for extra transportation,

b) Whether figures are presented after deducting extra equipment installed in the meanwhile,

c) Whether estimated figures (i.e., those which are not recorded from calibrated instruments) are presented in *italics,*

d) Whether the specific energy figures are on feed-to-plant basis, or based on the dominant product/ a few of the more important products (plain sum or weighted sum; basis of weighting),

e) Policy/ vision/ strategy adopted by the unit: Annex #

f) Analysis of energy scenario of the unit in terms of 'Energy Cost/Total Cost' and 'Energy Cost/ Variable Cost': Annex #

g) Energy audits undertaken equipment wise if any: Annex #

h) Energy conservation activities undertaken in the plants: Annex #

i) Statement on instruments (existing/ working/ calibration): Annex #

j) Improvements made in instrumentation if any: Annex #

k) Further disclosures if any: Annex #

An Accounting Framework for Presenting Specific Energy and Energy Conservation Figures for an Industrial Plant

A presentation of energy conservation figures in an accounting frame work such as the above may be jointly certified by the energy conservation and the finance department as representing a fair picture of energy conservation at the unit before the same is submitted for further action elsewhere.

Each plant should be asked to evolve its own practices of presenting energy conservation figures and declare these in the space for disclosures and certification. A panel of engineers and accountants may also be considered to help evolve practices suitable to each plant. The practices are generally likely to be different for each industrial plant.

Not the Last Word

No doubt, with adoption and use, the accounting framework shall evolve over time. And save valuable hours of executive altercation.

Tyranny of Errors

My Own Suffering is the Beginning

I suffer from poor memory. Consequently I suffer in the hands of my merciless examiners, superiors, juniors and my good friends and colleagues. I attempt to hide behind the adage "It is human to err". This article is an offshoot from this deficiency of mine.

INEFFICIENCY, ERRORS, MISTAKES, blunders, mischief and crime: these words describe progressively higher degrees of human failure in performance. By way of illustration, we present the following table:

Characteristic		**Exhibited by**
Inefficiency	:	Bulk machinery; petrol and diesel cars (~40% efficiency); transformers, motors and generators (80-99% efficiency; electric incandescent lamps (5% upwards) etc.
Errors	:	Instruments; wrist watch, thermometers, energy meters, flow meters, pressure gauges (0.1 to 10% errors) etc.
Mistakes	:	Human being. "It is human to err".
Blunders	:	Very conspicuous mistakes.
Mischief	:	Creativity, love, play, friendship exhibit mischief as an element.
Crime	:	Harmful acts.

It is our intention to argue that whereas crimes must be punished with exemplary severity, mischief should be tolerated with a degree of nobility of spirit, blunders must not be repeated, mistakes must be accepted as a part of human frailty, and errors and inefficiencies must be lived with as necessary evils of the material world.

When a First Class engineer is recruited by a management it is known clearly that the candidate would not have received 100% marks in the qualifying examination.

The degree certificate of the engineer is the equivalent of the name plate capacity of equipment. This declares his nominal performance. If he has 70% marks, he is expected to perform up to 70% of perfection. He is rated to fail up to 30% times. He will commit 30% errors in judgment, planning and execution. Even with experience the error shall not be zero.

A professor in literature, when writing a letter, occasionally corrects his spellings and sentences. Can you hold this against his scholarship? No. This type of mistakes is a part of the intellectual machinery, common to all human beings. You cannot deny the professor his scholarship on account of these mistakes. Just as a machine handling bulk power has a certain irreducible amount of inefficiency, just as an instrument handling information has a certain irreducible amount of error as illustrated in the above table, so has a human being, handling a job, a certain amount of irreducible amount of mistakes committed by him.

It is a myth that if everybody commits mistakes, the world will fall apart. It will not as can be seen from what follows:

In an hierarchical organization, say, X reports to Y who, in turn, reports to Z and nominally, each one is, say, 70% efficient in handling his allotted job. In an ordinary scenario, as follows, mistakes actually go on reducing as the job moves from person to person.

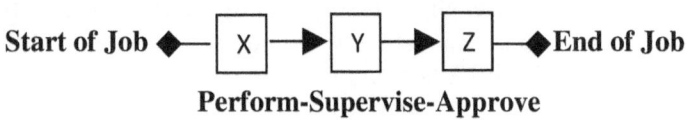

Perform-Supervise-Approve

The job is performed by X, supervised by Y and, finally, approved by Z. At each stage, assume, 30% mistakes are made by X, Y and Z respectively. A mistake will go past both X and Y if and only if X commits the mistake (30% of the times) and Y over looks the same (30% of the times). This means 30% of 30%, i.e. 9% (say, 10%) of the mistakes will go past Y's supervision. And the job will be properly executed with up to 90% correctness.

At the stage of approval, Z will commit his share of 30% mistakes. 30% of 10% is 3% ; i.e., 3% mistake will got past the approval stage. The job will be executed with 97% correctness. The world will not fall apart because of human mistakes.

In the real world, engineering processes, where proportional automatic control concept is applied, the control function operates on the basis of errors. The whole system, i.e., the plant along with the automatic control system, can function beautifully within say, about 2 to 3% error.

Mistakes will be committed by man. Punishment for mistakes will only make him hide his mistakes or make him seek to put the blame on someone else. System will lose the benefit of the consequent corrective effort. What is worse, the hidden errors will, in any case, do their invisible damage in course of time.

Sri Krishna says to Arjuna in the Gita: (BG 18:48)

*"Sahajam Karma Kaunteya sadosam api na tyajet
Sarbarmbha hi doshena dhumenagniribavritah".*

"Every initiative has some blemish in it like fire is covered up with smoke: nonetheless, Arjuna! Perform your duty! As work, according to ones call of duty, must be undertaken".

In the language of this essay we may translate the same as follows: "Notwithstanding the possibility of committing mistakes, Arjuna! Perform, because there is no work that can be done without some mistake"

Amen.

Time Estimation of a Job

How I Got the Basic Idea

Since my college days I used to make a number of inventions (except for one, none other was of any use to any one). I found I could not meet the target date as promised to my professors despite hard work, careful estimation and follow up. I observed, later, that I used take normally three times as much time as pledged. So I look to multiplying the estimated ideal time with three before committing completion time to my professors. In my industrial career I finally understood the underlying reason.

How a Job Proceeds

A JOB NORMALLY INVOLVES several agencies none of which works at 100% efficiency. Besides, in many organizations various formalities and the prevailing work culture pre-empts a proactive effort which ultimately results in a series operation. That is to say, one agency does a part of the work and only after that another agency shall take up another part.

For the purpose of this essay let us assume that every agency has an overall efficiency of 70% . In what follows 'T' is the ideal time estimated for the job (which may tacitly but wrongly assume 100% efficiency on the part of the various agencies involved in a job).

As an example, take a job where only one agency is involved, like, for example, a motor re-winding shop. The following would happen:

$\eta(A) = 70\%$

JOB ⟶ [A] ⟶ 70% JOB in 'T': 100% JOB in 1.5 T

Another example

Say, the engineering department has to complete one stage of a work using (A) departmental facilities before entrusting the balance to the agency of (B) a local contractor. Then the following would happen: (Series operation):

$\eta(A) = 70\%$ $\eta(B) = 70\%$

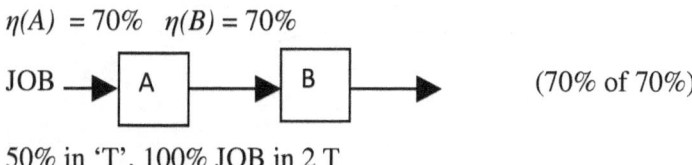

(70% of 70%)

50% in 'T', 100% JOB in 2 T

Generalizing, we would infer as follows:

AGENCIES	% of job in T	100% job in
A	70%	1.5T
A &B	70% *70% = 50%	2.0T
A,B&C	70% *70% *70% = 35%	3.0T
A,B,C & D	25%	4.0T

Real life frequently presents 3 agencies for a sizeable job:

Example A: 1. Initiator, 2. Workshop, 3. Contractor

Example B: 1. Organizer, 2. Middleman, 3. Service outlet

We can represent the above calculations in a graphical form (see the Figure below :)

Time Estimation of a Job

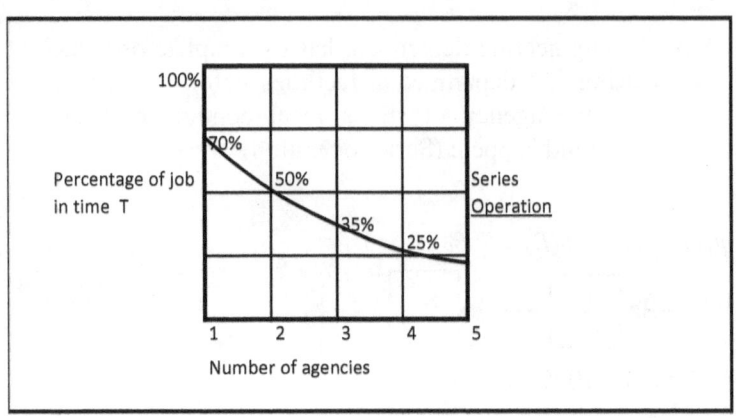

A solution is visible from the above discussion:

Parallel Operation of a Job

Say a manager can effectively call upon any or all of a number of facilities to do a particular job:

Example C: 1) Workshop
               2) Contractor

Example D: 1) Automobile section.
               2) Private taxi service.

Example E: 1) Major contractor
               2) Local contractor

In such cases the job will get done faster. This is known from experience. Let us make a model:

Time Estimation of a Job

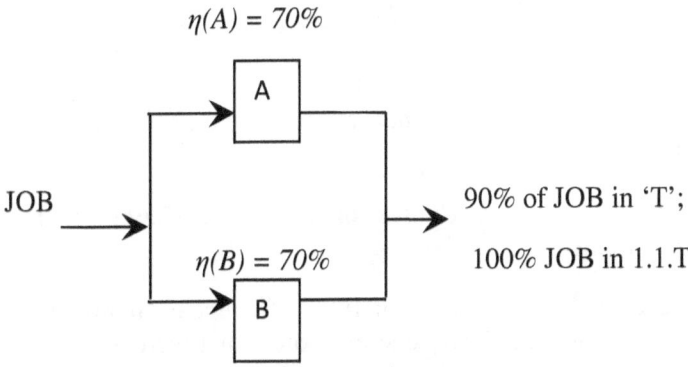

(*Calculation:* 100% - [30% x 30%] = 90%)

Say the Auto section can give a vehicle 70% of the times on demand. If on a specific demand they cannot provide a vehicle (i.e. 30% times) then there is the taxi service (also at 70% availability) to fall back upon. A vehicle will NOT be available if both the arrangements fail, i.e., 30% * 30% =10% times a vehicle will not be available. In this way a parallel arrangement will ensure availability of vehicles up to 90%. If yet another taxi service is also recruited then the NON availability will be only 10% * 30* = 3% i.e. 97% times a vehicle will be available.

If we generalize the above example we see, graphically, the following:

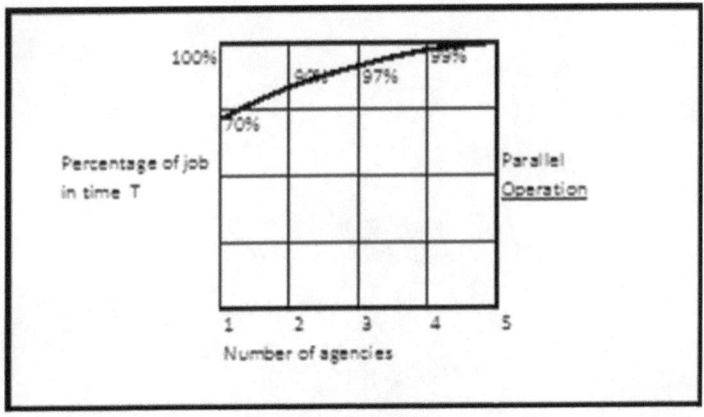

Analytically (as different from the graphical) η^N is the net percentage efficiency in *series operation* of the N number of agencies involved, and [100 $-(100-\eta)^N$] is the net percentage efficiency in *parallel operation* of the N number of agencies involved.

The Reality behind Claims of 100% Efficiency in Job Management

Whenever there are claims of 100% efficiencies in any of the departments in an industry one may know that there are hidden parallel arrangements like, for example, rate contractor for motor rewinding shop, or, local labor contractor or a taxi service, or, the very best, a willing boss readily okaying various proposal to engage various service providers.

Industrial Organization Modeled as a Pump–Pipeline Network

How I Got This Idea
This chapter may be read along with the next one: $P = Q \times R$ and an industrial organization. The ideas of both the chapters are concurrent.

Basic Idea

SEVERAL ASPECTS OF an industrial organization can be intuitively modeled after a pump-pipe line network (hereafter called, simply, 'pumping network'). This model, like any other model, physical or intuitive, should not be stretched too far. It is important to remember that every model is a mental object and, as such, can be stretched or assaulted up to a limit beyond which it will breakdown; that is to say, every mental object, too, has an yield point (static loading) and a fatigue limit (dynamic loading) just like any material object. It is common experience that good and sound ideas are, indeed, destroyed every day around us by being stretched too far.

Work flows in an industry just like water flows in a pipe line network. An executive in his area of responsibility can be linked to a motor-pump set. An executive derives authority from the organizational charter and drives his department just as a motor drives a pump while deriving electric power from a power source.

Industrial Organization Modeled as a Pump–Pipeline Network

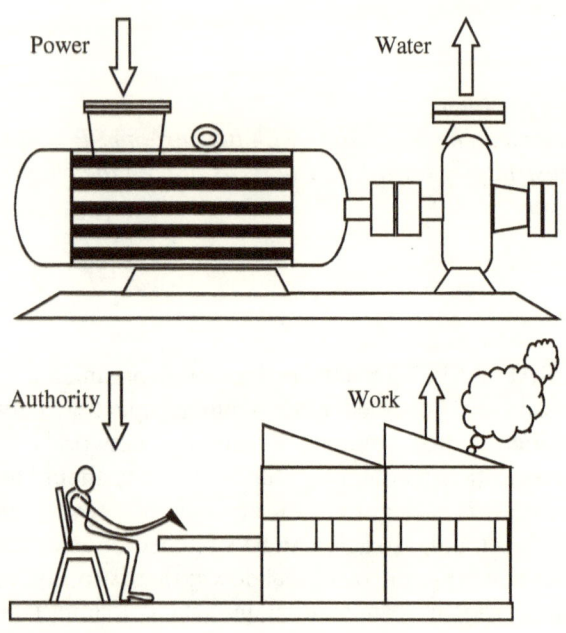

Let us ask a few questions and see if the pumping network model can come up with correct answers, which can satisfactorily explain the workings of an industrial organization. After all, that is the very purpose of toiling with a scientific model. Say, the downstream pumps do not work. The flow will, consequently, reduce. If the top pump tries to effect the designed flow everywhere, and is not adequately protected from this suicidal effort, it will generate all the pressure. This will lead to great pressure near the top pump, cause leakage, heating and mechanical damage. Same thing for a top executive in an organization where down the line executives do not or cannot (owing to distrust) exercise their authority in order to 'discharge' (mark this word) their responsibility. And just like idling pumps such executives can become drags on an organization instead of being drivers.

In a pumping network if more and more valves are put for the purpose of monitoring and control, then, beyond a point, more obstruction will result in the path of flow leading to further pressure losses and leakages which, in turn, will lead to further power requirement. Same thing happens to the flow of job if more and more top management interference comes into play. A time will come when, in order to get over the confusion and frustration, the top man has to assume total load of supervision, engineering and management with easily foreseeable results.

We note, there are three different flows associated with a pumping network:
- (a) Water in pipe lines:
- (b) Electric power in cables;
- (c) Information in instrumentation and control

Same is true about an industrial organization:

1. Job flow ⇔ water flow;
2. Authority ⇔ electric power;
3. Information ⇔ information.

A Diversion

Consider the flow of information. What is the pressure that drives the flow of information?

It is hoped the pump-pipeline model as proposed above shall prove helpful to grasp the workings of an organization intuitively.

P = Q x R and an Industrial Organization

How I Got This Idea

I don't think I can really say how I got this idea. It was a stroke of intuition. Perhaps unconsciously I looked for the pressure behind the flow of work and, then, logically, got around to various components of the net pressure of work.

Pump-Pipeline Model for an Organization

MATHEMATICALLY an organization can be modeled after a pump-pipe line network where flow of work resembles the flow of fluid. (One might, if one likes, model an organization after an electrical network with voltage sources and impedances.) Simple and useful consequences appear from such an analysis. This may help remove many of the common ills that plague the organizations and improve productivity which is, largely flow of work.

We propose the following law for organizations relating the flow of work to various pressures motivating flow of work and resistances impeding the same:

Organizations $P = Q * R$ (Law)

Pressure   =   *Flow rate*   *   *Resistance to the*
of work   *of jobs*   *flow of jobs*

It is a very simple relationship among quantities of flow of jobs and resembles similar relationships in various branches of science and engineering. For example:

Hydraulics: $\quad\quad\quad P = Q * R$
Pressure $\quad$ = Flow rate of water * Resistance in pipe line;

Electricity: $\quad\quad\quad V = I * R$
Voltage $\quad$ = $\quad$ Electric current * Conductor Resistance;

Thermodynamics: $\quad\quad T = Q * R$
Temperature = $\quad$ Flow rate of heat * Thermal Resistance.

We notice that, generally, where there is a flow there is a pressure driving the flow against the resistance in the path of the flow. (As an interesting diversion, think of the train of thoughts incessantly going on in your mind, involuntarily. Discover the 'pressure' driving the 'flow of thoughts'. Obviously there is a pressure because the thinking does get accelerated to a high pitch when you are under great pressure. Besides, it takes a great deal of exertion to tame or stem the flow of thought. This may possibly start you on a path of *Raja yoga*.)

One sees that our daily language is a rich reservoir of intuitive models and these can be gainfully employed for understanding life. Art precedes, and succeeds science.

Pressure $\quad$ = $\quad$ *Flow rate* $\quad$ * $\quad$ *Resistance to the*
of work $\quad\quad\quad\quad$ *of jobs* $\quad\quad\quad\quad$ *flow of jobs:*

$\quad P = Q * R \quad\quad\quad\quad\quad\quad$ (Law)

Let us look at the components of each term of the above relationship:

Pressure of work:

$\quad\quad \Sigma Pi = P1 + P2 + P3 +$ etc.; where

P1 = Authority, P2 = Target, P3 = Search for excellence etc.

Flow rate of 'work;
$\quad\quad$ So many tons of products per shift;

So many inspections conducted per day,

So many vehicles repaired per week, etc.

In every case we see a quantity referred to a fixed period of time.

Resistance to work:

This is the most interesting part. In fact this is the very animal we are after to improve productivity in the organization. This has got several static and dynamic components. First, the static components:

$R1$ = demoralized workforce,

$R2$ = un-streamlined procedures,

$R3$ = aging of plant and machinery,

$R4$ = propensity for negative decisions,

$R5$ = executives who dodge work,

$R6$ = workers who dodge work, etc.

Next the dynamic components:

$R7$ = Opposition to change (this is similar to 'inertia of mass' or 'inductance of electric inductor')

$R8$ = hoarding instinct (this is similar to a 'reservoir' or an 'electric capacitor'; the tendency to hoard unnecessarily large amounts of materials in various sub stores in the fear of a possible shortage; this will prevent optimum availability of materials to all concerned at the right time and right place and, thus, shall hamper the flow of work).

We, further, see that the stated law yields

$Q = P/R$

this implies that reduction of resistance to work shall raise the output.

It shall call for great sensitivity to discern all the various components of resistances, both static and dynamic components, and predict their combined effect to trouble shoot an organization which is not doing well in terms of job flow rate.

What Next?

Set up management consultancies. Complicate an obviously simple idea. Make money.

A Structural Limitation of Descriptive Language: One Dimensionality

How I Got This Idea

I think, I suddenly realized one day, that an entire book can indeed be written down in a single long straight line. Much the same is true of the spoken language as one word follows the other in a strict sequence. This feature of one dimensionality constitutes an important geometrical structure of the descriptive languages used by the mankind. Various consequences of this limitation of one-dimensionality of the descriptive languages and various attempts to overcome the same are discussed.

Basic Role of Language

THE BASIC ROLE OF LANGUAGE is to transmit a mental image from Mr. A to Mr. B. We can model the process thus:

| Mental image of A | *Language* → | Mental image of B |

One can use the above mentioned model to identify several limitations of language inherent in its role as a vehicle of mental image.

Language is an artificial creation of the human mind. It is still developing. Every year dozens of words are created, borrowed and used with different connotations, and modifications. Our tacit assumption that language can express everything is not correct. One notices how a person falters, pauses, hesitates and then corrects himself, while giving words to an important thought or feeling for the first time.

Language introduces errors in the information it is supposed to transmit. The totality of the distortion in the transmission of signals is called NOISE by the communications engineer. All information carrying devices like the radio, the T.V., the computer are also subject to NOISE; this is also true about the language as a medium of transmission of mental imagery. This is one reason why we correct our statements, and sometimes entirely scratch and rewrite, letters, essays and poems.

Limitation of One-Dimensionality of Descriptive (Verbal and Written) Language

A further limitation of the descriptive language arises from the fact that it has a one dimensional structure. Anything that can be put on a very long straight line is inherently one dimensional. You can write a book on a thin strip of paper of perhaps ten kilometers in length. One word comes after another and is followed by exactly one more word. In contrast to the descriptive language, a map is two dimensional and an idol is three dimensional.

The one dimensionality of structure of descriptive language makes it difficult to express our experiences which are multi-dimensional. To one who has never seen an aero plane, it is difficult to describe the experience of a flight. However, aided with pictures, the job is easier. It is easiest if one gets hold of a wooden model of an aero plane. However, this is a trivial example. The really beautiful and powerful examples of successful attempts by man at surmounting this limitation of one dimensionality of descriptive language are: poetry, drama, song, dance, mathematics, humor, jokes and the like. For example when one says, "A bird in hand is worth two in bush" one is neither speaking about a bird nor a bush. If one actually tries to describe the idea to another in a plain manner, one will take full five minutes.

To give yet another beautiful example, consider Walter de la Mare's poem "Napoleon":

A Structural Limitation of Descriptive Language: One Dimensionality

"What is this world O soldiers! It is I. I,
this incessant snow, this northern sky.
Soldiers! This solitude through which we go is I."

This, if we take it purely literally, is pure nonsense. Nevertheless the poet has been able to express high spirituality by overcoming the limitation of one dimensionality of descriptive language.

Of humor I shall give no examples.

I will give two more examples to show how man has attempted successfully to overcome the limitation of one dimensionality of language. One is pure music, instrumental, which are without words. The other is what the physicist Dr. Douglas R. Hofstadter calls breaking the mind of logic in his celebrated book *GÖDEL, EISCHER, BACH: AN ENTERNAL GOLDEN BRAID*. The example is a Koan. Koan is an anecdote form used by Zen Buddhists to shock pupils into great revelations. I have lifted, with due apologies, a Koan, a commentary and a poem by Mumon a great Zen sage, and a further commentary by Dr. Hofstadter from his book.

"Koan:

A monk asked Nansen: "Is there a teaching no master ever taught before"?

Nansen said "yes, there is".

"What is it?" asked the monk.

Nansen replied. "It is not mind, it is not Buddha, it is not things"

Munon's commentary:

Old Nansen gave away his treasure words. He must have been greatly upset.

A Structural Limitation of Descriptive Language: One Dimensionality

Munon's Poem:

Nansen was too kind and lost his treasure. Truly, words have no power. Even though the mountain becomes the sea, words cannot open another's mind.

In this poem Munon seems to be saying something very central to Zen, and not making idiotic statements. Curiously, however, the poem is self-referential and thus it is a comment not only on Nansen's words, but also on its own ineffectiveness. This type of paradox is quite characteristic of Zen. It is an attempt to "break the mind of logic". As well as transcending the structural limitation of one dimensionality of descriptive language.

It seems better to leave off with the beautiful excerpts from Douglas Hofstadter's.

Certificates

Certificates

Dr. M. Bhattacharyya
M. Sc. (Engg.) (B.H.U.), Ph. D. (U.K.) MIEEE (U.S.A.),
Associate MIEE (U.K.), C. Engg., MIE (Ind.), MIIST (Ind.)
Professor of Electrical Engg., (Electric Drives & Traction)

Phone : 54291/374
Department of Electrical Engineering
Institute of Technology
Banaras Hindu University
Varanasi-221005 (INDIA)
Date 29th April, 1980

TO WHOM IT MAY CONCERN

I have pleasure in certifying that Sri Lingraj Patnaik, appeared in the final semester of the Third Year Electrical Engineering Examination of the Institute of Technology in April, 1980.

Sri Patnaik is an extremely innovative young man, and his originality in thinking and his ability to apply his knowledge to practical use has been amply demonstrated in this short period through the development of a prototype non-conventional transverse pedal bi-cycle, with improved performance. Sri Patnaik's efforts have been highly acclaimed at different levels of the University and outside.

I wish him a very successful life and further hope that he would continue to pursue his creative ideas.

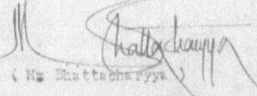

(M. Bhattacharyya)

Certificates

Institute of Technology
Banaras Hindu University
Varanasi - 221005
MODEX-80

This is to certify that Sri Lingaraj Patnaik of IV-Year Electrical Engg. Class has been awarded Consolation prize for his model Vertical Pedal Bicycle in Educational category exhibited in the Engineering Models Exhibition held from 8th to 13th Oct. 1980.

Director
Institute of Technology

Convenor
MODEX-80

Institute of Technology
Banaras Hindu University
Varanasi - 221005
MODEX-80

This is to certify that Sri L. Patnaik of IV-Year Electrical Engg. Class has been awarded Second prize for his model Wave Boat in Advanced Technology category exhibited in the Engineering Models Exhibition held from 8th to 13th Oct. 1980.

Director
Institute of Technology

Convenor
MODEX-80

Certificates

Dr. S. Suresh Kumar
Dy. General Manager (TS & QA)

Indian Rare Earths Ltd.,
Orissa Sands Complex
P.O: Matikhalo – 761 045
Dist:Ganjam, ORISSA, INDIA
☎: +91-6811-63510
e-mail:sureshkumarire@yahoo.co.in

December 3, 2001

TO WHOMSOEVER THIS MAY CONCERN

INVENTION OF WAVE PUMP

I know Shri Lingaraj Patnaik, B.Tech (Electrical), BHU, for the last 18 years who is presently working as Manager (Electrical) in charge of Energy Conservation, OSCOM, IRE Ltd., Matikhalo- 761045, Orissa, INDIA. He has pursued several original ideas and innovations with keen interest.

During 1988- 89 Shri Patnaik developed a prototype 'WAVE PUMP'. He was fundamentally inspired by the observation that fish wave their bodies and tails to propel themselves while swimming.

On 25th August 1988 his prototype was demonstrated in the Technical Services Laboratory, OSCOM, IRE Ltd. The 'WAVE PUMP' generated a maximum of 3 meter water head at zero discharge. At lesser water head the discharge was approximately 6 gallons per minute and was quite smooth and regular. The principle of pumping water by the new principle of wave motion of a suspended plate was conclusively demonstrated.

I wish him all success.

Dr. S. Suresh Kumar
DGM (TS & QA)

ATOMIC ENERGY CENTRAL SCHOOL, OSCOM

Affiliated to CBSE Code No. 15356, School No. 8392
I R E HOUSING COMPLEX, PO.MATIKHALO : PIN : 761045, DIST. : GANJAM (ORISSA)

Ph. 06811-63890-95
Extn. 154

Ref. No. AECS/05 Com/2003/4456

Date: 28/2/03

To

Shri Lingaraj Pattnaik,
Sr. Manager,
IRE,
OSCOM

Sir,

Thank you for contributing the concept for the preparation of the project "Transverse Wave Propulsion in Bird's flight."

I am glad to inform you that the above said project has been selected for participating in Inter AECS Science Exhibition to be held on 13th & 14th August, 2003 at Manuguru.

With regards

(GOPINATHAN A P S)
I/C., PRINCIPAL

Certificates

Building Wound-Core Transformer

BY
LINGARAJ PATNAIK, M.I.E.

ABSTRACT:

A method of building transformer... illustrated. ... will have obvious theoretical and practical ... of m... and repairing will be reduced. The manufacturi... ly mechanised.

CONCEPT:
Building the Core

PLEASE refer in Fig. 1. Building the core can be accomplished with a mould having two straight sides and two curved sides. Avoiding sharp bends to the ... ... internal stresses ... ... on a horizontal turn... ... cined to achieve the two ... ... of the core as precisely ... ... tilting rollers ... ... used for this purpose. ... ... tion tight bracings must be used to maintain the ...

Conductor Winding

Please refer to Fig. 2. For winding the conductors, a splittable mould will be introduced about one straight side of the core. It can be held floating with the help of various rollers. The mould may have, at one or both the circular ends, means to attach a belt, a chain or a gear to drive it. Then suitable conductors can be wound upon it. This mould will be left behind in the transformer. When the occasion arises for repairing the transformer, the windings can be unwound with the help of this mould ... ... the other windings of either one phase or three phase transformers ...

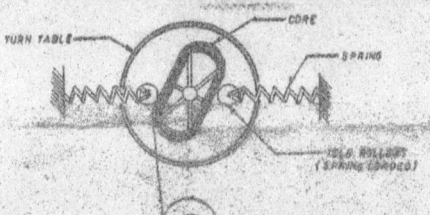

Certificates

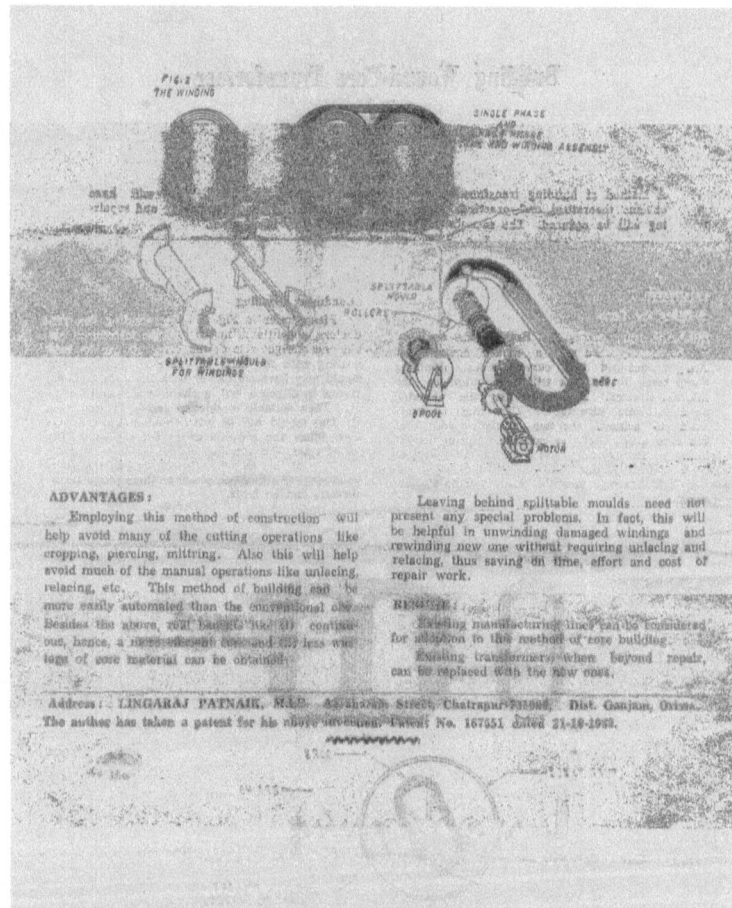

ADVANTAGES:

Employing this method of construction will help avoid many of the cutting operations like cropping, piercing, mittring. Also this will help avoid much of the manual operations like unlacing, relacing, etc. This method of building and be more easily automated than the conventional one. Besides the above, real benefits like (i) continuous, hence, a measurement core and (ii) less wastage of core material can be obtained.

Leaving behind splittable moulds need not present any special problems. In fact, this will be helpful in unwinding damaged windings and rewinding new one without requiring unlacing and relacing, thus saving on time, effort and cost of repair work.

REMARKS:

Existing manufacturing lines can be considered for adoption to this method of core building.

Existing transformers, when beyond repair, can be replaced with the new ones.

Address: LINGARAJ PATNAIK, M.I.E., Aiginia Street, Chatrapur-761020, Dist. Ganjam, Orissa. The author has taken a patent for his above invention. Patent No. 167551 dated 21-10-1989.

Reprinted from : **ELECTRICAL INDIA** — 30th JUNE, 1993
Proprietors : Chary Publications, Unit No. 33, Vaibhav Industrial Estate, Near Telecom Factory, Sion-Trombay Road, Deonar, Bombay-400 088. Phone : 551 82 54

Certificates

GOVERNMENT OF INDIA
THE PATENT OFFICE

A No. 14794

No. 167551 of 21 - 10 - 19 86

WHEREAS LINGARAJ PATNAIK (INDIAN), AGRAHARAM STREET, P.O. CHATRAPUR, DIST. GANJAM, ORISSA, an Indian national,

has/have declared that he is/they are in possession of an invention for core and winding assembly for transformers,

and that he is/they are the true and first inventor(s) thereof [struck through] and that he is/they are entitled to a patent for the said invention, having regard to the provisions of the Patents Act, 1970 and that there is no objection to the grant of a patent to him/them;

And whereas he has/they have by an application requested that a patent may be granted to him/them for the said invention;

And whereas he has/they have by and in his/their complete specification particularly described and ascertained the nature of the said invention and the manner in which the same is to be performed;

Now these presents that the abovesaid applicant(s) (including his/their legal representative(s) and assignee(s) or any of them) shall, subject to the provisions of the Patents Act, 1970 and the conditions specified in section 47 of the said Act, and to the conditions and provisions specified by any other law for the time being in force, have the exclusive privilege of [struck through]

/using or exercising his invention for core and winding assembly for transformers,

in India, for a term of fourteen years from the twenty first day of October 19 86 and of authorising any other person to do so, subject to the conditions that the validity of this patent is not guaranteed and that the fees prescribed for the continuance of this patent are duly paid.

In witness whereof, the Controller has caused this patent to be sealed as of the twenty first day of October 19 86.

Controller of Patents

15th Nov, 1991/23 Krtk,1913 (Saka)
Date of Sealing

Note: The fees for renewal of this patent, if it is to be maintained, will fall due on _____ day of _____ 19 ____, and on the same day in every year thereafter.

Certificates

Solution to Nuisance Tripping of Plant Lighting from Fault in Appliances

By
Lingaraj Patnaik MIE

Nuisance tripping of plant lighting circuits protected by Earth Leakage Circuit Breakers is a nightmare for electrical maintenance people especially when a portion of the plant lighting trips for no better reason than an earth leakage or earth fault in, say, a plug socket outlet.

Whereas safety is achieved upto 100%, circuit availability is reduced below acceptable levels. And this, in many instances, leads to by-passing of the ELCBs by the maintenance people.

The trouble arises because the same ELCB which potects the luminaire circuits also protects the power socket outlets meant for appliances.

Whereas luminaires do not develop earth faults frequently the same is not true for appliances hooked to power socket outlets where frequent earth faults do take place.

The solution to the above problems could be in the following directions:

1) All plug sockets and appliances should be fed from independent MCB DBs to be designated as 'Appliance MCB DBs.' Only luminaire circuits should be fed from 'Luminaire MCB DBs'. If this is done plant lighting shall not suffer nuisance tripping owing to appliances.

2) If however, the same MCB DB is to be used for both luminaires and appliances then 2 phases should be used for luminaires and the 3rd phase should be reserved for plug sockets and appliances only. Of course plug & sockets should be kept in boards separate from luminaire switch boards.

3) 100 mA/300 mA ELCB may be used for the 'luminaire MCB DBs' or the 'Luminaire phases' (2 out of 3 phases) and 30 mA ELCB for the 'Appliance MCB DBs' or the 'appliance phase' (3rd phase).

This will ensure fire hazard protection (100/300mA) for the luminaire circuits and personal safety (30 mA) for appliances circuits, and, thus, we feel, will provide the optimum performance for the combined availability and safety aspects.

PUBLISHED!

Reprinted from : **ELECTRICAL INDIA** — 31st AUGUST, 1993

Proprietors : Chary Publications, Unit No. 33   Vaibhav Industrial Estate, Near Telecom Factory, Sion-Trombay Road, Deonar, Bombay-400 088. Phone : 551 82 54

FAULTS IN APPLIANCES LEAD TO NUISANCE TRIPPING OF LIGHTS & FANS FROM TRIPPING OF ELCB. AN ENGINEERING METHOD TO OBVIATE THE DIFFICULTY

Certificates

Institute of Advanced Computer & Research
(Managed by Rabindranath Educational Trust, Rayagada)
(Approved by AICTE, Govt. of Orissa and Affiliated to Berhampur University)

Ref..................   Date: 29.03.2003

Certified that Prof./Dr./Mr./Ms. _Lingaraj Patnaik_ has participated at the 30th Annual Conference of Orissa Mathematical Society (OMS) and the National Conference on "Role of Mathematics in Engineering & Technology", sponsored by CSIR, New Delhi and presented the paper entitled _A new approach to the Conditions for analyticity of complex functions._ during 22nd-23rd of February, 2003.

(Prof. B. B. Panda)
Principal,
IACR, Rayagada
Institute of Advanced Computer
& Research, Rayagada.

Address: Prajukti Vihar, Rayagada - 765 002 (Orissa), Ph.(06856) 225850,225246,225270,225827 Fax: (06856) - 225246
E-mail: iacr_mca@hotmail.com, iacr_engg@yahoo.com   Web Site: www.iacrcollege.com

INDEX

Aircraft, 16ff
Analytical function, 82f
Arjuna, 119
Atom bomb, 99ff

Bicycle, 9
Birds, 12ff
Boat, 12ff

Carbon, 101ff
Charge density, 44
Chemical equation, 94
Complex, 57ff, 78ff, 101
Conservative force field, 81

Electromagnetic, 32, 44ff, 47ff, 52, 59
Energy, 38ff, 59ff
Energy conservation, 107ff
Energy, kinetic, potential, 38ff, 89
Environment, 98ff
Error, 29ff, 56ff, 107ff, 114ff, 131

Heat, 97ff, 124, 127
Hiroshima, 97ff
Holfstadter, Douglas R. 132f
Hydraulics, 127

Imaginary, 58, 80f, 91
Industrial, 107ff, 118, 123ff, 126
Inefficiency, 115
Information, 101, 115, 125, 131
Integral theorem, 79ff
Inverse square Law, 42ff

Job, 115ff

Koan, 132
Krishna, 116

Language, 127, 130ff
Laplacian, 83, 86f
Length contraction, 56f, 74
Lorentz transformation, 34f, 57, 65f, 69

Matrix, 47ff, 70f, 76ff
Maxwell, 29ff, 42, 44ff
Mischief, 114
Motor, 13, 16, 101, 114, 118, 122, 123

Newton's laws, 12f, 29ff, 38ff
Noise, 19, 99, 131
Organization, 103, 115, 118, 123f, 126ff

Index

Poem, 131ff

Population, 99ff

Potential, 38ff, 53f, 81, 83, 88ff

Power, 14ff, 101ff, 107, 115, 123ff

Pump, 13ff, 99, 123ff, 126

Radiative Forcing, 97

Relativity, 33ff, 56ff, 60, 65f

Resistance, 89, 102, 126ff

Scale, 56ff, 76ff, 103ff

Shear force, 43

Specific Energy, 107ff

Stratosphere, 99

Streamline, 42, 128

Tensor, 47ff

Zen, 132

www.ingramcontent.com/pod-product-compliance
Lightning Source LLC
Chambersburg PA
CBHW020917180526
45163CB00007B/2766